Forest Sovereignty

Plants and Animals
Interdisciplinary Approaches

Series Editors
Professor Susan McHugh, University of New England
Professor Patrícia Vieira, University of Coimbra

Vol. 1

Gregory F. Tague

Forest Sovereignty

Wildlife Sustainability and Ethics

PETER LANG

Oxford · Berlin · Bruxelles · Chennai · Lausanne · New York

Bibliographic information published by the Deutsche Nationalbibliothek
The German National Library lists this publication in the German National Bibliography; detailed bibliographic data is available on the Internet at http://dnb.d-nb.de.

A catalogue record for this book is available from the British Library.

Library of Congress Control Number: 2025023697

Cover image © Cover art by Karolina Jacks-Tague. Reproduced with permission.
Cover design by Peter Lang Group AG

ISSN 2753-3255
ISBN 978-1-80374-955-6 (print)
ISBN 978-1-80374-956-3 (ePDF)
ISBN 978-1-80374-957-0 (ePub)

DOI 10.3726/b22703

Published by Peter Lang Ltd, Oxford, United Kingdom
info@peterlang.com - www.peterlang.com

This publication has been peer reviewed.

Where others see God, I see Gaia
In memory of my parents, Fran and Joe
And again, for Fredericka and Karolina

Contents

Acknowledgments

This work, the result of analyzing and synthesizing words and ideas from many people, began early in 2022 when I considered comments from others about the broader environmental implications of my book, *An Ape Ethic and the Question of Personhood*. Fredericka A. Jacks through many conversations helped ferry some concepts out of darkness. I'm grateful for feedback from first and then anonymous peer readers and help from editors including Susan McHugh, Patrícia Vieira, and Laurel Plapp. Special thanks to John Dilyard for comments on the manuscript, to Todd B. Adams for assistance regarding Native Americans, and to Luis E. Banegas for help with voices from the Global South. My thanks to Ionut Untea, an editor at the American Philosophical Association website, for publishing my essay "The Wildlife State of Gaia," July 2024, related to the ideas here without quite using the words of this book. There are too many other people to mention individually, and I thank them for their dedication in advocating for animal rights and environmental preservation.

This book is dedicated to my parents, who raised me to be intellectually inquisitive, liberally fair-minded, and diplomatically honest.

Original cover art has been provided by Karolina Jacks-Tague.

Preface and Summary

Human societies are still wrestling with the best form of government in spite of political philosophy from Plato to the present. Following Socrates, *The Republic* is concerned with how society's conflicting interests, whether oligarchy or democracy, allow one to flourish (Cornford 1975, p. xv). Human history shows there are empires that rise and fall, but there's no reason their declines should disintegrate the fabric of earth's biomes. If justice in Plato's formula is the distribution of fair rights and the fulfillment of honest obligations, we find that formation among forest flora and fauna where the physical behavior and social conduct of one positively affects many others. Humankind is mostly centered on physical desires; paradoxically, the earth's nonhuman organisms are almost beyond space and time as elements of a huge biosphere. University professors, corporate executives, and politicians are keen to talk about sustainability, but it's almost always in anthropocentric terms and economically driven. We should turn, rather, to forest wildlife and understand how they are true ecological sustainers working forest lands along with flora. Humans have devised an array of ethical systems but mostly treat forests unfairly. Instead, forest bio-networks evince a self-sustaining ethic of care to all inhabitants.

No political theory explains the successful self-determining government we find in the forest state of Gaia. Some might see Gaia as an ideal society where, in line with a biological reading of Aristotle's *Politics*, the community evolves with individual and group adaptations for all to live well (Hadas 1950, p. 154). Nature's realm has already solved the problems of fairness and equity with which human communities are still grappling. This book offers a theory of forest sovereignty based on nature's interspecies mutual cooperation manifesting in ecological engineering for planetary health. In arguing for forest sovereignty, it will be useful to compare and contrast key political thinkers, with the understanding that Gaia operates as an effective system reflecting salient parts of human political thought. The purpose of this book is to evaluate social practices related to nature, and such an inquiry is normative since it raises questions of ethics and rightful actions. Political philosophy for the past few hundred years, with a focus almost exclusively on the human, has failed nature from which we evolved and on which we are dependent. Yet the essentials of political philosophy across thinkers discussed hereinafter can be applied to forest sovereignty; where political philosophy has failed in the human realm (e.g., property distribution), it has succeeded with forest life. Individuals, politicians, corporations, and states can possess, transfer, sell, and legislate forest lands. By that logic of property ownership, the flora and fauna of wild lands are the inheritors, owners, and managers of forests, which in my definition hereafter includes bordering wetlands and grasslands.

Human power, whether political, social, or technological has been wielded against nature. Misguided use of force to shape nature in egregious ways for humans is never justified. Like other organisms, we evolved brains to make decisions. Yet we choose to destroy forests (a fact) that ensure the longevity of earth's climate health (a value). John Reid and Thomas Lovejoy (2022) argue these points with data. Noted as early as 1967 in *The Theory of Island Biogeography* by Robert McArthur and E. O. Wilson, there's a correlation between forest size and biodiversity. Fragmentation spawns a cascading effect of species loss, which in turn diminishes the potency of a forest's carbon capture and ecological engineering abilities to clean air and water. In contrast to groups of Indigenous people in forests, industrialized humans have used forest flora and fauna for their own needs and desires. That's a factual statement rife with moral implications. As John Christman (2002) notes, a key question in political philosophy asks what justifies state power over people. An overwhelming theme of political philosophy, intones Colin Farrelly (2004, p. x), deals with making choices about ideal "social arrangements," emphasizing only the human element. The operational system in forests is not really power over others. That's a human perspective. Rather, forest wildlife and plants are free in cyclical movements of life and decay until, without their consent, they and their habitats are yoked under limitations imposed by humans.

Political philosophers have misunderstood the balances of power and shared spaces in nature by assuming that the control of natural resources is best for humans. Liberal philosophy emphasizes autonomy of the individual in a free society, but much of individualism is tied to tradition, culture, and social norms. While evolution by natural selection is not progressive and can't look to the future in its randomness, it seems more liberal than any traditional conservative ideology that bases the present and future on the past. Forest wildlife seems liberal in its ability to mutate and adapt via individual selection and yet, it's tied to a past since evolution can only improve what's already there. The forest state of Gaia is a community of vital individuals whose everyday activities regarding survival, reproduction, and feeding behaviors represent selection benefiting the entire planet. As historian William McNeill (1976) suggests, lethal epidemics changed the course of history in war and peace. By analogy, modern humans act as a scourge on nature through profit-oriented misadventures of deforestation fueling economic development from industries as wide-ranging as cattle farming to suburban sprawl.

While not the deadliest global pandemic, what is known as Covid-19 killed over 7 million people about four years after its known outbreak. As a zoonotic disease, Covid-19 likely spread from a wet market where animals are slaughtered and sold on demand. This meat-centric practice sounds like something from the Middle Ages. Humans as the supposed wisest and most successful species tend to center everything in themselves, anthropocentrism. The secret of human success has been contriving ways and means to use nature to our advantage, satisfying our perceived needs and desires. Another virus of human concoction has been spreading silently and unnoticeably for hundreds of years: deforestation and the corresponding elimination of wetlands and grasslands. The reach of global warming from deforestation promotes unclean air and impure water, but it's within our power to reverse course if voting citizens and nation-states exercise a plant-centered environmental willpower. We are overdue for political change regarding the damages committed. Western laws are mostly anthropocentric and benefit human economies. Instead, laws could be amended to include the needs of wild forest flora, fauna, and fungi; existing regulations could be rigorously enforced with consideration for all life forms. While the concerns of Indigenous people are present, the emphasis of this book is on nonhuman nature. Incidentally, the designation Indigenous is used loosely considering the vast diversity of people and their cultures; generally, the term applies to those whose views of the natural world are close to the concept of Gaia presented in this discussion.

The point is to show how political theory developed by humans can also be applied to a theory for the self-governing state of Gaia. The focus on forest science and early political thinkers expands outward in subsequent chapters and concludes with ecological ethics and Indigenous voices. Starting this project, I assumed some political notions would help explain forest dominion. While the main idea seems to hover mostly around the liberalism of a welfare state, there are elements of utilitarian and libertarian theory. There are many forests, but all are connected and working in unison at different times of the year on various parts of earth for climate stability. Applying political theory to forests is to argue for the sovereignty of global Gaia that regulates the planet's climate.

Therefore, without disenfranchising Indigenous residents, Gaia should be left alone as a sovereign state to continue performing its vital functions of cleansing the atmosphere of carbon pollutions. We know this is feasible since many nations are beginning to establish larger forests. Even more contiguous lands should be isolated, as suggested by the efforts of legendary biologist E. O. Wilson and his project called *Half-Earth* (2016), where 50 percent of land would be left to wildlife. John Reid and Thomas Lovejoy (2022) convincingly argue that mega forests and their cores must be saved for planetary health, yet like Wilson they offer no political theory to do so.

I'm offering competing political philosophies to show that there's no single theory of justice to fit the forest except for Gaia as a self-governing state. As political philosopher Will Kymlicka (2002, p. 211) points out, justice is a way for society to evaluate its practices, not simply reinforce them. I'll try to reconcile some differences to demonstrate what's crucial for carbon dioxide reduction and a lessening of the baneful effects of climate change. Corporate agriculture is now buying carbon credits from small farmers. That's not a solution, since corporations will continue to deforest land in the Global South for profit by farming beef cattle, soy or corn feed for livestock, and biofuels for export (Keupper, et al., 2019). In new profit schemes, companies are planting trees in deforested areas, with no guarantee of ecosystem growth, only to sell carbon credits to polluters with no promise of reduced fossil fuel emissions (Andreoni 2024). Richard Schmitt (2009, p. 172) flatly states that compared to humans, animals are inferior and exist for our use.

That's a dangerous attitude which needs to be changed. In a short section of his book (2009, pp. 180–83), Schmitt asks if animals have rights. The practice of abusing another body logically suggests it has some degree of sovereignty. Humans are cognitively, emotionally, and physically related to many other species. Schmitt goes on to note how the suffering of animals should afford them "some" rights. These academic implications are not constructive to the interrelated wildlife, forest, and climate crisis we now face. On a philosophical level, it's considered undignified to take away someone's freedom, yet that belief does not seem to apply to how we treat forest animals. There's overlap between what I call forest sovereignty and the rights of nature movement. Wildlife, plants, and trees can have legal standing; nature has inherent value beyond any utility for modern humans; natural rights can ensure conservation; Indigenous cultures, of which there are many varieties, recognize the sanctity of nature with practices that ensure sustainability. If there's a difference, forest sovereignty, in addition, asks that flora, fauna, and fungi remain abundant and alone to continue servicing global, atmospheric health in an effort to preserve Gaia.

The takeaway from my argument is that we are dependent on woodland, wetland, and grassland wildlife and plants for our future existence and need to treat flora, fauna, and fungi not only better but also as our political equals in the realm of justice. Evolution has selected for an animal ethos geared to sustainable stewardship of the forest in an overall group effect of mutual interactions and dependencies. Vincent Zonca (2023) argues that any idea of symbiosis should not include mutualistic notions. A focus, rather, should be on an indefinite relationship that contains competitive cooperation epitomized, for example, in lichen composed of fungus (earth) / alga (sea) and laden with microbes so the whole is almost void of singular identity. Especially on a microbial level, symbiosis is a long-term relationship between two different species that, according to Lynn Margulis (1998), can be a driver of evolution. Mutualism implies beneficial but not necessarily symbiotic interactions between organisms. However we look at the situation, the real question is to what degree modern humans are in a symbiotic or even mutualistic relationship with what we call nature.

Mutually interrelated, animals in the wild along with plants, trees, and microorganisms, service the forests and holistically maintain equilibrium in a sustainable environment (Mayr 1997, p. 19). It's popular nowadays for industry executives to apply the term ecosystem to their businesses. That's inappropriate since ecology deals with the natural world. Contrary to any ecological thinking, business entrepreneurs have now constructed a huge artificial intelligence barge in international waters calling for national sovereignty to avoid government oversight (Del Complex 2023). In a similar way, the word biodiversity is bandied as if a magic wand. In truth, the key for Gaia is how all of the biodiverse species in an ecosystem perform interlocking and recycling functions of survival. Biodiversity is valuable for its net results. The biodiverse facility of ecosystem engineering to sustain habitats is an evolved ethic of the land that humans should respect. Going further afield from any idea of a kin relationship between wildlife and forests (Tague 2020), creatures in the wild exhibit a value agency. This does not have to be conscious. Rather, look at the total results. Life forms from microorganisms to mega fauna work together in ways (agency) that ecologically sustain habitats for their flourishing (value).

Among developing and developed societies there could be an engagement of ethical care toward forests with expansive rewilding and reforestation, nonintrusive land management, and woodland rejuvenation. In tandem, local governments could invest effort in educating young citizens with voting power about the causes and consequences of, and remedies for, climate change from deforestation. According to a Pew study (Tyson & Kennedy 2023), over 60 percent of Americans, and even more young people, believe climate change will increase though few expect to make significant sacrifices. Climate change cannot be denied or ignored. Renowned naturalist David Attenborough (2020, p. 175) says that deforestation is the "emblem of our dominance" over nature. When we cut down trees or eradicate wetlands or grasslands, surface biodiversity suffers, which ultimately cascades into a negative ripple effect for the local ecosystem, the biome, and eventually the biosphere. We are witnessing this crisis in real time through human-induced climate change while we have a good part of the solution at hand in forest sovereignty.

The Virus of Human Climate Sovereignty

Concerning climate change, there's political polarization beneath the attitude of human sovereignty over planet earth. Adherence to a cohesive social identity can actually prevent people from accepting fact-based science (Hoffman 2015). Americans who resist accepting climate change are less likely to care for people or animals affected in distant locations (Hart & Nisbet 2011). Science and facts can, paradoxically, sometimes reinforce existing ideology. Attitude consistency without any flexibility to new, credible ideas creates an illusory truth. One example of this adherence to illusion is found among people in the U.S. who denied the outcome of the 2020 presidential election in spite of evidence and court cases. Misinformation bias exists in terms of animal rights, forest preservation, and climate change. A study indicates that confirmation bias and attitude confidence exist among many people who search the internet to reinforce what they already believe (Schweiger & Cress 2019). False knowledge consists of repeating caustic misinformation creating an illusory truth (Hassan & Barber 2021). The only correction to this ideological contagion is continued influence and successful debiasing (Lewandowsky, et al. 2012).

The more often human destruction of the plant world, there will be less biodiversity of significant organic matter (Eshaghi Rad 2018). Paul Raskin, et al. (2002) encouraged scientists and leaders over 20 years ago to take action about climate change, building on the then older idea of sustainable development. The virus was visible, and the calls were made, but with little action the disease spread. As a benchmark, we might want to consider if we averted impoverishing nature, and thus human well-being, since those times in what's now a historical transition. With the rise of the personal PC and internet in the 1980s and the collapse of the Soviet Union in the late 1990s, says Raskin, et al. (2002), we entered a rapid period of global connectivity and expansion with a wedge growing between those who envisioned catastrophe and those who hoped for transformation. We also entered an age where misinformation is easily spread across various media.

A Modest Proposal

Two related solutions might mitigate the current climate crisis, but human willpower is lacking: 1. a vegan culture, as I argue in *The Vegan Evolution* (2022); 2. forest sovereignty, as I argue here, building on my book *An Ape Ethic* (2020), which focuses on apes as ecosystem engineers. Other solutions, though not as final as numbers 1 and 2, could include the following. Promoting laws like those in the U.K. ("ancient woodlands") or Australia ("old-growth forest") to protect lands; Endorsing statutory language that protects forests, as seen in the constitution of New York State regarding the Adirondacks and the Catskill mountains; Promoting an international forest corps service component for young people (financed by governments) to educate youth about the importance of forests since the future belongs to them. These solutions do not require technology but cultural variations in attitudes and whom we elect.

Let's tender an offer: Local activists and young people could form a citizen panel, parliament, or community trust including Indigenous persons, guided and informed by legal, science, and ethics experts to inform their community about the urgency of forest protection; then, the citizen stakeholders could influence local, regional, state, and national elected officials to take action initiating legal forms of forest sovereignty by means of conservation, rewilding, or preservation. Specific ideas related to this proposal appear in the section Human and Animal Proximities in Chapter 4.

Coming Up Roses?

Some might say, why argue for forest and wildlife sovereignty with so many people suffering. Putting aside the truism that advocating for animal rights can also address wrongs to humans, we see how political philosophies in some parts of the world have mostly failed poor and marginalized people. Some will accuse me of naivete, of presenting a rosy picture of animals in nature. They will say I have no idea how creatures in the wild suffer miserably and sometimes die tragically in what Rachel Carson (2002, p. 249) calls "terrible intensity." I know all of this. What those critics don't see is that I am not focused on a species or groups of animals but the entirety of the forest world including contiguous wetlands and grasslands across continents. There are many moving parts to the energized world of Gaia, and certainly some individuals flourish more than others, but all live and die as nature intended. That's the point. When we look at the whole of Gaia corroded at the hands of industrialized people, the unvarnished truth is that humans are the deadly culprits, not any purported predators or parasites that evolved with all nature's life to balance populations and keep the energy supply pumping. The world of nature is not a chain of command; that's a human perspective generating images of predation. The picture of Gaia need not be pretty just because humans like things clean and shiny without decay; the dirt of this earth keeps us alive.

The Compass of This Book

On one hand, this book offers a critique of Enlightenment political philosophy from Thomas Hobbes and John Locke up to John Rawls and Robert Nozick including Michael Walzer; on the other hand, the book explores anthropocentric political philosophy to offer a theory of forest sovereignty using the very ideas from these seminal thinkers. With an ecological political theory in the foreground, maybe then better environmental legislation could follow. Typically, following Locke, political philosophers argue that what exists in nature becomes human property with added value once labored over by people. This is a Western, colonialist settler mentality that has fostered deforestation and wildlife degradation. The Introduction lays out basic ideas and themes related to the thesis. Chapter 1 familiarizes readers with basic concepts concerning the life of trees and other wildlife since forests are a central subject. Chapter 2 takes the scientific findings from Chapter 1 and aligns them with the political philosophy of Enlightenment philosophers. Then, in Chapter 3, the thinkers from Chapter 2 are positioned in line with twentieth-century political theory. Before pointing to the Conclusion, which attempts an ecological credo including Indigenous voices, Chapter 4 locates thinking from the foundational and more recent political philosophers in a broader discussion of justice. Readers will find that the chapters mature in a logical sequence. The book navigates itself from a controversial thesis about forest sovereignty to concepts about how rights and freedom can be applied to the preservation of nature for planetary health.

Introduction

Moral and Political Margins

Trees are a political issue pointing to three reasons why forest sovereignty is paramount: 1. Forest expansion is a feasible answer, as biologists suggest, to mitigate the climate crisis; 2. Trees as ultimate ecosystem engineers can be considered ethical subjects of a life like animals; 3. Wildlife works with flora, as fungi work with flora, to sustain our biosphere. Developers want denuded land in order to build sprawling housing developments. Corporate logging companies want trees for profit. Hunters would prefer fewer trees to open up spaces for deer. Industrial agriculture would like flattened land in the Amazon or elsewhere to raise beef cattle and the grain needed to feed them. Conservationists want to preserve trees for aesthetic reasons and to fend off global warming. Ecologists see forests as vast biomes of biodiversity crucial to the health and well-being of the planet. An editorial by Guy Robinson (2024) in a special issue of *Geography and Sustainability* makes clear that even with technology-enhanced farming, deforestation is not sound land management. There's global potential for natural regeneration in deforested areas, says Brooke Williams, et al. (2024). Farming research shows the importance of remnant trees as ecosystem soil servicers in agricultural areas (Ding & Eldridge 2024). For all that, people are often controlled by illusions. Faulty ideas can spread like a virus with dangerous beliefs like technocentric solutions to climate change or agriscience forest loss. Writer Ben Rawlence (2022, p. 160) is correct, and the import of this book says, that forests have "geopolitical relevance" across the globe.

Mick Smith (2011) writes about the presumptuous dangers of political sovereignty over the natural world, a form of biopolitics after Michel Foucault (1991). Nations assume dominion over nature as if it's a pool of resources, he says. Smith does not espouse natural sovereignty because that grants nature too much human-like political power. Rather, he advocates for ecological ethics that recognizes the primacy of place in nonhuman others. The problem is that we need a tool, whether legal or political, to protect and conserve wildlife, plants, etc., and why an argument for forest sovereignty seems appropriate. Even Smith agrees that many political theorists from Plato onward see the world as governable, but the forests, wetlands, and grasslands of Gaia are self-governing unless impeded by human politics and machines. In this respect, Smith does not consider the bountiful eco-engineering capacities of forests and wildlife. Random elements in nature allow for the adaptation and evolution of multifarious forms and ecological niches comprising a complex totality. Jakob von Uexküll (2010) established Umwelt theory, or the life-world of subjective realities where organisms fit into a defined environment completely based on receptor cues to physical sensations and movements in response to perception of time, space, temperature, etc. In one Umwelt an organism is a subject; in another's Umwelt the same organism is an object. Humans strike to impose their restricting sense of order on the openness of nature resulting in terrible outcomes for nonhuman forms and the climate. Smith (pp. 159, 163) says sovereignty as political dominance must be eliminated. Granting sovereignty to nature's forests is not an act of authority. Sovereignty will empower the salubrious energies of plants and wildlife to flourish unabated as they have adapted to fit into their self-sustaining world that fortifies all forms of life across ecosystems.

As Thomas Kuhn asks in *The Structure of Scientific Revolutions* (2012), what is the pressing problem that needs to be solved above all others? The political theorists examined, from Hobbes and Locke to Rawls and Nozick, wish to alter society in some way. My aim is to explore all of these views in terms of forest sovereignty and to compare and contrast a forest state of Gaia with human conceptions of any ideal state. Gaia is transnational. In the balance, we'll see that political philosophy has ignored wildlife, but yet many of its ideas can be applied to forest organisms. Later, I'll talk about Gaia as a multinational state. We need to consider forest sovereignty from an ethical as much as from a scientific perspective. Christine Gudorf and James Huchingson (2003) see environmental concerns as central to ethics. Systems of morality should not consider only humans since the natural world is entitled to rights. Without question there is currently conflict between humans and forest flora, fauna, and fungi; this conflict must be ethically resolved for the sake of all life. We are not morally obligated to preserve museums that house stuffed animals. We are ethically obligated to preserve living forest wildlife upon which the health of earth depends. There's an ethical claim in saying forests have incalculable value because their intrinsic qualities of growth, metamorphosis, and decay are instrumental in creating and maintaining any number of ecosystems that constitute earth's biosphere.

In the final paragraph of *On the Origin of Species* (2003), Charles Darwin talks about a tangled bank as an orderly system supporting multiple species and structures. In nature, the secret to success is not dominance but interdependence. In a similar fashion, early humans evolved in and then emerged from forest and savanna ecosystems with multiple, archetypal identities. In nature, not counting small populations of Indigenous people, modern humans have forgotten their origins and have thought only how to benefit themselves to the great detriment of other life forms. As Marcy Norton (2024) explains, in early modern times some Indigenous people among the Americas tamed wild animals as friendly companions but allowed them to roam free without exploitation. European conquerors, Norton suggests, did not quite grasp that humans and animals were part of and shared the same ecosystem. Colonizers were accustomed to raising animals as livestock to be slaughtered and eaten. Eventually, according to Mackenzie Cooley (2022), European settlers commercially farmed animals in the Americas for food and hides, which led to negative changes in the landscape and environment. Meanwhile, flora and fauna working for survival independently and with mutual cooperation encourage forest growth for the benefit of all forms of life via climate health. Like animals, we should understand the vital importance of living with and protecting, not eliminating, forests and contingent environs.

No one has a crystal ball to see the near future, but we require ecosystem solidarity, which through cultural evolution could ameliorate the worst effects of the climate virus and the destructive attitudes that still deny it's happening. Cultural evolution consists of socially learned and quickly acquired habits, knowledge, beliefs, and capabilities (Hannon & Lewens 2014; Tague 2022, ch. 5). Any view of the world in climate crisis is not opinionated but shared by a consensus of scientists (Oreskes 2018). Cultural elasticity can be a potent means of change. As an update to his earlier publication, Raskin (2021) speaks in terms of communal disease when he suggests that earth's climate meltdown from industrial and consumerist activity will cause "cultural polarization," "social disparity," and "resource depletion," all of which are now present and must be cured by electing political leaders with the will for saving the crucial bio-network of flora, fauna, and fungi by endorsing a forest state of Gaia. Toward the closing decades of the twentieth century many industrial countries increased reforestation, but this transition could be illusory since so much has already been lost and other areas are under threat. One analysis concerned with food production for developed nations (Waggoner & Ausubel 2001) focuses only on croplands, with an optimistic prediction of a small increase of forest cover. However, a year after that prediction was made Global Forest Watch revealed a forest loss of 2.4 percent from 2001 to 2021, with some areas like the U.S. losing 16 percent.

To start, let's offer some guiding principles outlined in this Introduction and returned to in the following chapters:

1. Many people would fundamentally agree to life, liberty, and a safe environment.
2. Political theory has mainly failed regarding some human and most forest security.
3. Political theory for humans could be used to understand nonhuman societies.
4. Forests have a form of justice based in a shared biology of mutualism.

5. Industrial nations have yet to respect fully the global benefits of healthy forests.
6. Applying political theory and philosophy to forests is a step toward their sovereignty and therefore the well-being and safety of all life forms.

The Awful Truth

Between 2000 and 2016, there has been a 10 percent decimation or loss of intact forests because of drilling and mining north of and logging south of the equator (Reid & Lovejoy 2022, p. 8). From 1977 to 2008, there was a 66 percent decrease in extant forest resulting not only in tree and biodiversity loss but also in severe woodland fragmentation (Bianchi & Haig 2013). If not left to replenish themselves, in 25 years these forests will be gone. Intervention, especially by local people trained not to damage these areas, can protect forests to expand fragments leading to rewilding. Widespread deforestation, stemming from factors like overpopulation, suburban development, animal agriculture, mining, logging, etc., is a leading cause of climate change (Geist & Lambin 2002; Wuebbles, et al. 2017). From 2000 to 2005, forest cover loss from human or natural occurrences totaled over 1 million square kilometers. This loss equates to 3 percent of total forest cover of 32.6 million square kilometers in the year 2000. Most loss happened in the boreal biome and then the humid and dry tropics followed by temperate areas from Canada, the U.S., and Brazil (Hansen, et al. 2010). Threats of deforestation and resulting biophysical effects to already compromised ecosystems are escalating at about 10 million hectares per year in a report (Almond, et al. 2022) that highlights forest loss since plant life, trees, canopy cover, etc., link together biodiversity, climate stability, fresh water, and woodland food resources.

Increased animal agricultural production, given market fluctuations, can occur with less deforestation pending government and corporate policies that efficiently utilize land. However, cattle ranching and the lands needed to feed livestock are major drivers of deforestation (Macedo, et al. 2012). Regarding corporate animal agriculture, over 75 percent of global soy is used as livestock feed in meat and dairy farming, with some going to biofuels, but only 7 percent consumed by humans (Ritchie & Roser 2021; X. Song, et al., 2021). A study with quantified models shows that a quick decline of animal agriculture will stabilize greenhouse gas for decades and offset CO_2 levels (Eisen & Brown 2022). Global demands for biofuels and grains to feed animal agriculture for the meat industry account for forest dieback and loss of "ecological services" like atmospheric water circulation and the creation of carbon capture savanna (Nepstad, et al. 2008). To reduce global warming, forests must be preserved from human threats. The conversion of croplands to grow industry biofuels, for example, will increase greenhouse gases through emissions and land use change (Searchinger, et al. 2008). This loss of tree coverage and the mismanagement of forests, woodland, wetlands, and grasslands (Noss, et al. 2021) is so pervasive and ingrained into the culture of industrial nations, one wonders how it can be stopped and the damages reversed.

In one study providing a tree census (Crowther, et al. 2015), it was determined that there are over 3 trillion trees globally, with most in tropical regions, followed by boreal and temperate areas. These authors estimate tree loss at about 15 billion per year, meaning that forest Gaia has lost over 45 percent of its trees since the dawn of human civilization. Other scientists (Bastin, et al. 2019) estimate that 4.4 billion hectares of restoration tree cover is possible, providing about 205 gigatons of carbon storage. At the same time, climate change with hotter air and dryer soils will hinder efforts of tree restoration with anticipated shrinkage of millions of hectares of canopy cover in the tropics. Research (Osuri, et al. 2020) strongly suggests that regeneration of natural, multispecies forests, rather than promoting commercial tree monocultures, is the best way to mitigate effects of climate change through carbon dioxide capture. The message is, let it grow wild and leave it alone.

These few data points should alarm readers and help them realize they have power for change in whom they elect and what they teach their young.

Philosophy of Trees

In his book *The Life of Plants*, philosopher Emanuele Coccia (2019, p. 4) laments how plants are often disregarded by people in general and by zoocentric scientists in particular where "superiority" is granted by humans to the animal and not to the plant world. Coccia also says, pertinent to this discussion, that plants don't evoke our empathy or moral sense like some animals in our anthropocentric views. Nonetheless, unlike animals who can move, fly, burrow, hide, hibernate, etc., plants experience intimacy with the natural world we've mostly lost. Additionally, unlike most forms of life that must feed from others, plants are self-sufficient in their total being; they are primary. Almost all other life is secondary and dependent on plants for nourishment and oxygen. Coccia says that whereas most organisms adapt to their environments, plants have shaped the world in their ability to keep growing. He calls for a "philosophy of nature" above humanism since the physical world doesn't simply precede but confronts and joins us in genesis and metamorphosis.

My argument can sound anthropocentric: healthy earth for humans. On the contrary, animals and the forest in which they live have their own rights. Rather than utilitarianism (via Peter Singer) or deontology (via Tom Regan), forest flora, fauna, and fungi have inherent rights to live and function in their own evolved and evolving manner in the habitats they've carved as ecological niches. Philosophers Sue Donaldson and Will Kymlicka (2013) would agree. A niche consists of environmental ranges for an organism and helps it relate to and survive in an environment; a niche can also evolve because of an organism's genetic variation (Holt 2014, p. 288). Population size increases or decreases because human alterations precipitate "cascading changes" across plant, animal, and fungal organisms in the community (Strauss 2014, p. 302). Compressing forest, wetland, and grassland areas through human activity increases competition and population density leading to adverse living conditions. High levels of cooperation across a spectrum of forest species creates alliances favoring adaptation, whereas a decrease of species biodiversity leads to an impoverished ecosystem (Strassmann & Queller 2014, pp. 675–76). Species interact and affect niche size and variation. Humans do not support or supply these survival rights that have been earned over millions of years. Invoking biocentrism, Christine Gudorf and James Huchingson (2003, p. 9) say that all living organisms including microbes have a "teleological center" that generates movements good for them, conscious or not. Biocentrism sees humans as interdependent with nature, but does not really go beyond individuals to ecosystems as does Aldo Leopold (1949). According to Leopold, we have a land ethic because we feel for forests, wetlands, and grasslands. Ecosystems as a whole containing all forms of life can suffer and are, in some oblique manner, sentient.

In this way, the nearsightedness of political philosophy is like an opaque lens between humans and the environment, blurring what should be seen clearly. Vegetation is the measure of all things, not humans. Climate is what unifies all living beings, and photosynthesizing plants are the most integral components of the entire atmospheric "mixture" touching and unifying all in a continual genesis, says Coccia (2019, p. 27). He sees matter as fluid, evident in the many metamorphoses of living beings over time that are fully immersed in the world. Coccia exemplifies my argument by saying we inhabit the atmosphere and do not simply occupy the earth (2019, p. 35). We are in both a physical and a fluid space, the latter of which we have treated with neglect for hundreds of years. Plants are great modifiers of surface and air, the epitome of metamorphosis in the relations they have with other beings and inanimate objects. Whereas plants "merge" into a world (Coccia 2019, p. 41), we do not. At least we have not tried to do so in any holistic manner. Some people are now coming to this realization by imagining how city buildings can be reengineered to dwell in nature. One example is the city of Medellín, Colombia.

Coccia (2019) does not think plants fit any model of niche construction and says Darwin's book on worms, advancing a niche construction concept, goes against natural selection. Niche construction, he says, does not allow for the kind of worldly immersion we see in plants. These theoretical perspectives are not incompatible depending on the species and the environment. Research by Maria Przybylska, et al. (2024), for example, reveals how plants use enzymes to modify root areas with nitrogen and microbes ensuring their health and preparing for a new generation. The niche is a construction of habitat so as not only to define but also to share space. Individuals vary and compete in populations with the selection and heritability of survival and reproductive success for some, not all (Mayr 1991, p. 76). Darwinian selection is qualitative, not quantitative as per Malthus. Simplistically put, genes respond to and in some cases influence a demarcated environment, as with worms. In the forest, no living organism lives in a walled enclosure like humans. Plants exert considerably more positive consequence on the world than humans. This is because most humans live only in artificial environments and have essentially lost any mutually healthy correspondence with nature. Coccia goes on to note how, as early as 1772, Joseph Priestly deduced that plants manufacture healthy air. A few years later, Jan Ingenhousz pinpoints this action in plant leaves as photosynthesis (Coccia 2019, pp. 44–5). All animal life is, as he says, *dependent* upon and *determined* by the oxygenation activity of plants.

Thus, what we call adaptation is complicated by an organism's immersion in a total environment, not in an enclosure, and its relationship to the many other organisms similarly situated, especially those in soils. Privatizing and enclosing "common space" to extract resources for capitalist gains and consumer wants is at the root of forest degradation and without cultural change will only hasten the climate crisis (Burdon 2015, p. 6). We need lots of trees to breathe, and forests consist not only of trees but also of biodiverse organisms that interact with and sustain life in one shared atmosphere. As per Vladimir Vernadsky, says Coccia (2019), the atmosphere is not separate from but is a crucial spatial facet shared by all living forms. Forests create an extension of earth where material life is malleable and grows. Trees and plants breathe for and into animals like us, but because we don't see the blending of essential particles, we take this process of equilibrium for granted. In many ways, plant life has created animal life. Some people might say we are like animals or suggest there's an animal part in us. Both of those assertions and assumptions are false. We *are* animals and like every living creature we need green space with clean air and clear water.

Fossil evidence shows that roots are a later plant development with differing functions over time (Coccia 2019, p. 78). Roots fill the soil with information and communicate with other plants, mineral elements, and microorganisms. Among roots is a nether world as important and vitally connected to life on earth's surface. As Coccia notes, in this way plants inhabit multiple spheres and not necessarily one niche; they are darkly earthbound and brightly airborne simultaneously. Humans could learn through plants that inhabiting two worlds is not mutually exclusive but unified. We need to understand that we belong in a natural world though we drive cars and live in houses. Coccia talks about plants having two bodies, and so we too must recognize presence in our world and dependence on theirs. To imagine we don't need forests is like saying we don't need skin. Just as trees and plants need the sun to feed, and as animals need plants to eat, we need as much forest cover as possible to hydrate and cleanse the air we breathe. Still, we must pollute less. Just as the sun is center to our universe, forests are central to a wholesome planet earth. Trees transpire water, which creates rain and impacts wind flows; so, in addition to capturing carbon, they act as global cooling mechanisms.

Setting Boundaries

In Brazil during the 1830s, Darwin perceived the "dangers of deforestation" (Preston 2022, p. 77). Then, in 1867, an American scientist cautioned Wisconsin state legislators that deforestation would negatively affect the climate (Reid & Lovejoy 2022, p. 165). Russian scientists in 1961 warned of the harmful transformative effects of "climate change," their expression in a 1962 publication after an earlier conference (Rawlence 2022, p. 87). They warned of polar ice destruction and the loss of healthy air circulation. In 1965, the U.S. White House issued a report that directly connects technology in an affluent society with a range of pollutants, especially carbon dioxide. The statement (Report 1965) discusses imminent effects of greenhouse gases from the burning of fossil fuels, like the melting of the Antarctic Ice Cap, warming and rising sea levels, etc. Daniel Moynihan (1969) followed up on this report with a White House memorandum warning of atmospheric carbon dioxide increases from fossil fuels. The scientific evidence and some political will about global warming were evident in the middle of the twentieth century when there would have been time to plan collective action and educate a voting public.

By 1977, Exxon was aware of the harmful effects of burning fossil fuels (Hall 2015). An assessment of oil company records from 1977 to 2003 demonstrates that industry-paid fossil fuel scientists accurately predicted, and as was subsequently verified independently, the "potentially catastrophic" global warming effects of burning oil, gas, and coal (Supran, et al. 2023). Public statements by the oil corporations hid and denied their own findings. Scientists and executives of the oil, gas, and coal industries were well-aware of the damages and dangers of anthropogenic greenhouse warming back into the 1950s and well into the 1980s. Few heeded these warnings and little was done. Corporate meat and dairy industries have infiltrated academia so that professors and their affiliated research centers are financially and ethically appropriated to avowedly minimize the impact of animal agriculture on climate change (Morris & Jacquet 2024). Management consulting giant McKinsey & Company completed a report pointing to how corporations could "restore natural capital" (Aminetzah, et al. 2022).

The ruling consensus among many but not all politicians and corporate officials prefers business as usual. One can't help but wonder if the audience for the McKinsey study is ethical individuals or executives concerned with public image and bottom line. So-called "natural capital" could be construed as forest timber for wood products, animal agriculture of livestock as food, or the monetization of minerals or water. For corporations, whether pharmaceutical companies digging up plants in the Amazon or cattle ranchers razing pristine lands, nature has become a cost/benefit analysis on a balance sheet. One report (Bakker, et al. 2021) that is business oriented offers a goal of transformative action in areas from energy, transportation, food, water, and sanitation. While anthropocentric, with a focus on helping businesses survive, the report is sensitive to the calamitous effects of deforestation on carbon storage and atmospheric hydration only because of impacts on commercial enterprise. Economic, social, and environmental boundaries are not separate. Regardless, in spite of the existing climate crisis and promises by countries to mitigate damage, leading oil, coal, and gas producers plan to ramp up fossil fuel extraction (SEI 2023), an indicator of corporate profits over climate stability and conservation of nature.

More ethically and practically, earth scientist John Rockström and nature journalist Mattias Klum (2015, p. 9) offer safeguards for the future, boundaries regarding climate change that must not be crossed. These include: 1. changes in climate; 2. diminishment in the ozone layer; 3. acidified marine life; 4. loss of forests; 5. depletion of fresh water; 6. biodiversity decrease; and 7. chemical air pollution. Many of these goals are tied to forest preservation. Without expansive, great forests to capture carbon with a concurrent elimination of fossil fuel production, there's no doubt the present climate crisis will worsen. Trees capture 50 to 75 percent of carbon emissions, says Ben Rawlence (2022, p. 35). Forests act as air purifiers, but that's not to justify air pollution. Forests cache huge amounts of fresh water. Working symbiotically with great forests, oceans produce about half of earth's oxygen, much shared with marine life (NOAA 2023). In addition to ocean habitats, forest ecosystems house extensive biodiversity as capable stewards of the environment. Many authors point out that facts alone will not be enough to advance change in the minds of some policy makers and an uninformed public. We require a change of culture that includes voices of Indigenous people who respect Gaia.

Through cultural evolution voting citizens can speak out and demonstrate how environmental concerns are entwined in social and economic outcomes. Since developed nations have used most of earth's resources to prosper and create wealth, they could be first to set aside and replant up to 50 percent of their forests or lands as forests, including restoration of wetlands and grasslands. Second, these nations could help developing countries conserve what's left of their forests with incentives to expand those areas by reducing monocrop plantation, carefully managing logging or mining, and eliminating cattle ranching and farming to feed livestock destined as human food. Plant-based diets are rich in proteins and much more efficient when done right. Feeding beef cattle requires more acreage and elimination of forest land than merely growing crops for human consumption (Stone 2010, p. 81). Greenhouse gases from animal agriculture are double that in plant-based food production (Xu, et al. 2021). Human deforestation opens large areas to more sunlight and promotes low-growth plants that in turn attract and help increase the population of grazers, like deer. This human activity prevents forest regeneration and creates an imbalance of wildlife by artificially boosting wolf populations who prey on deer and small mammals, not livestock.

There are few, if any, isolated economies today. Floods, droughts, powerful storms, etc., can occur locally or on some national level, but there are ripple effects globally to those crippled economies. Unless people act to elect nature-righteous legislators who will conscientiously fight climate change, severe weather events will only increase in speed and damage. Climate changes have occurred in the past, with quick and dramatic shifts of temperature and high doses of carbon dioxide shot into the air. Typically, though not always, via gradual Darwinian evolution, change is slow, and although many species would have died, some survived and evolved into variations and then new species. What's different now, Peter Wohlleben (2019, p. 166) suggests, is the rapidity with which the climate is changing, preventing tree species from adapting to a hotter, drier climate. As for adaptation, human development and other physical constraints don't permit trees to migrate naturally, which occurs as seeds fall from one tree to another locale expanding their growth boundaries over many years. Human interference withers soils, exhausts trees, and sets the stage for life-altering forest fires on a continental scale, already occurring. In some locations, natural wildfires are ecologically beneficial by destroying pests, opening areas for growth, and releasing seeds; this normal cycle is far different than the devastating human-made fires for corporate agriculture that foul the air and fuel climate change yet more (Gonzalez 2023b).

Many species evolved with trees in a dynamic dance of synergy; but trees are now warning us of impending danger by drifting north into areas previously secluded by ice. The boreal forest, says Ben Rawlence (2022), covers one-fifth of earth with one-half of trees, a huge pump of oxygen for the planet. The problem is that because of global warming, the boundary between temperate forests and the tundra is shrinking as trees advance quickly, not in Darwinian slow time. Not only is this alarming, but it also means earth loses the albedo effect of white ice reflecting the sun's rays. There are impacts on global air currents. A melting permafrost releases greenhouse gases accelerating the uncontrolled spiral of climate change. Trees far north are contributing to permafrost melt, releasing carbon stored there. Gaia's balance of cold/heat and ice/water between the tundra and the forest is upset, with bad consequences for life forms slow to adapt. This picture is complicated, Rawlence admits. Some research in Siberia and Alaska, Rawlence (2022, p. 126) goes on, suggests a stable tree line. Yet, the exposure of ground might have catastrophic and sudden consequences in terms of releasing greenhouse gases.

The virus of human-induced climate calamity has no limits. A burning forest in Indonesia or the Amazon scatters polluted air worldwide, affecting millions of people and animals. While there are separate nation-states, all share one environment, and unless that is truly acknowledged, the costs of the climate crisis will become more painful. In the words of R. J. W. Brienen, et al. (2015), there has been long-term decline of the Amazon carbon sink. On a collective basis, many humans are responsible for such tremendous loss. Reasonable steps include preserving existing forest cover and soil moisture, like those in the Amazon, Congo Basin, Southeast Asia, as well as boreal and temperate forests. Reforestation in these areas, including the reclamation of wetlands and grasslands, could be prioritized. Goals, as outlined by Alice Di Sacco, et al. (2020) include working cooperatively to reforest contingent wooded areas, coaxing regeneration naturally by not relying only on tree planting, and maximizing native plant biodiversity with rewilding.

Without great forests to stabilize the atmosphere and climate, the resilience of all weather systems can fail. Forest trees help rain clouds form because of tiny particles called terpenes they emit, says Peter Wohlleben (2019, pp. 159–60), which enable water droplets to gather together. Monoterpenes, says Ben Rawlence (2022, p. 21), emitted by Scots pine trees reflect sunrays from earth. Forests help form clouds, which offer a beneficial cooling effect to transport moisture and store water underground. As Rockström and Klum (2015) validate with clear data, local environmental losses in total have net global effects, and this is why conservation operations should begin with citizens voting on state and national levels, followed by legislators of conscience who can influence world leaders. Everything boils down to what future one wants, not what one needs today. Conservation is not simply the preservation of existing land but enabling plants and animals to engage in their adapted and evolving cultures necessary for ecosystem sustainability (Brakes, et al. 2021). Forest ecosystems are under increasing stress with poor biodiversity and deprived conservation

(Camia, et al. 2021). Climate change is a consideration for ecological boundaries of forests. Let's have more concern for the preservation of biodiversity with the cessation of illegal logging and a slowdown of rural development (European Court of Auditors). There has been some increase in wildlife, but challenges persist for the health and functionality of their global ecosystems (Ledger 2022).

Stable situations depend on positive feedbacks. The movements, feeding patterns, and other behaviors of forest organisms evolved in self-generating loops. If a line of the system is cut, fallout across the board is certain. By depleting natural resources, we only harm ourselves and other life forms necessary for planetary health. Rockström and Klum delineate this problem in detail with graphs and charts full of data, and they note that upsetting the climate affects food producers and consumers causing social unrest, as seen in Africa in 2010 (2015, p. 95). The sensible environmental boundaries Rockström and Klum propose negotiate opposition from extremes on the sides of endless resource extraction and apocalyptic soothsayers. This does not mean unlimited GDP growth. Instead, a primary focus could rest on recovering and restoring a healthy climate by establishing logical parameters to expand forests in all zones of the world. Local or national jobs could be created tied to forest, wetland, and grassland restoration and preservation since these places border on urban, suburban, rural, and farming areas. Cities become important in this picture because with a focus there on sustainable building and recycling of products and energy, rural areas can be left alone to rewild. Confirming this position, Caroline Fraser (2009, p. 87) notes that between 1960 and 2000 forestation in Europe increased about 70 percent as human populations moved from rural to urban areas.

By 2030 urban populations will have doubled with an additional 1.5 billion people (Schewenius 2014). Cities increase in size dramatically, and in biodiversity hotspots without foresight this growth could lead to irremediable plant and wildlife loss. By 2050, 70 percent of human populations will be concentrated in cities, raising new planning challenges about water use and sanitation, poverty, homelessness, energy consumption, etc. (Schewenius 2014). Not only could city expansion be designed around or include untouchable wild biodiversity, why not include an abundance of native trees, wetlands, pollinator plants and insects, reservoirs, aquifers, and community veggie farms to produce fresh and healthy protein-rich greens. A model green city-state is Singapore, where objectives and investments are made for social and environmental cohabitation. Children and teens can be involved in this cultural shift to engage them with nature. Human constructions and developments in cities could contribute to natural ecosystem services. Regenerative design (Benyus, et al. 2022) can occur through biomimicry to reduce city pollution effects on nature as well as habitat elimination and increase atmospheric, soil, and water cleansing cycles along with enhanced urban ecological communities. Redesigned cities could include biomimicry with green roofs, carbon sequestration cement, native plants, composting, and constructed wetlands. More so, cities could regenerate natural habitats by essentially enclosing built parks or natural grasslands, wetlands, or forest fragments with limited access. Invoking E. O. Wilson's notion of biophilia, one author (Cleveland 2014) emphasizes the connectedness between humans and nature with a need to include natural, functioning elements in urban design. Rather than manicured lawns that mimic the artificiality and pollution density of golf courses, some researchers (Tessler, et al. 2023) suggest suburban meadows to promote healthier microbial diversity and rehabilitate plant growth and habitat connectivity.

Through careful observation and documentation, Peter Wohlleben (2019) affirms how forests are in balance, from wolves and deer to microbes and plants. Modern humans have unwittingly upset that equilibrium, causing terrible effects, if not having deforested areas permanently. Wohlleben admits that humans still do not completely understand complex species relationships and should essentially leave forests to manage themselves. As Rockström and Klum (2015) would agree, technology and economics, with mindsets of GDP growth, need to be broken since they ignore the sustainable feedback systems of forests that replenish planet earth with clean air and water. Future-looking urban areas, say Rockström and Klum, include places like Copenhagen, Vesterbro; London, Beddington; Culemborg (Dutch), Eva Lanxmeer; and German Freiburg im Breisgau. These locales might serve as models for other cities globally. With pressure from constituents and the obvious effects of climate change, like severe weather and rising sea levels, city planners are taking heed.

China maximizes urban development and engages in reforestation initiatives but like other industrial nations can slip in some areas. For instance, China has converted a high rise building to pig farming (Zhao, et al. 2020; Guardian 2022; Wakabayashi & Fu 2023), but that's not an example of good urban zoning or sustainable farming. Indoor planting of vegetable foods would be greener and safer. The evidence is clear that we should move away from animal farming that wastes energy in storing, feeding, cleaning, slaughtering, packaging, and transporting butchered bodies (Machovina, et al. 2015; Springmann, et al. 2018; Clark, et al. 2020; Theurl, et al. 2020). As Waggoner and Ausubel (2001) conclude, by employing the best agricultural methods there can be more crop yields with fewer hectares, meaning less deforestation. Pig farming in a city raises concerns about local air and water pollution along with zoonotic disease, biosecurity risks, and other contaminations. In China, pig waste will supposedly be used to generate biogas, but that's no better than clearing lands to grow crops for biofuels or mining for "clean coal." Helmut Burkhardt and Richard

Westra (2008) state bluntly that agrofuels are environmentally detrimental, not sustainable, and should be replaced by solar or geothermal energy. China consumes 50 percent of global pork. This is inefficient food ecology, and meat is not the only source of protein. Pork has cultural significance to the Chinese, but attitudes can change with cultural evolution. Consider how cigarette smoking was a form of cultural acceptance in the West for many years until people realized its hazards. Boundaries regarding the cultural ecology of food can be beneficial, and city farming could include greens, not meats. Rockström and Klum (2015) argue for sustainable farming practices yet make no mention of a vegan culture but for a passing remark about lowering meat consumption. Moving to a vegan economy (Tague 2022) would reduce deforestation, meaning that we could replant those lands, thus opening areas polluted by cattle and other livestock farming for sustainable rewilding to replenish ecosystem feedback loops since broken.

Gudorf and Huchingson (2003) directly say that rewilding can occur as bulldozers manipulate earth close to its original form. That is not ecological engineering but cosmetic patching of destroyed areas. Animals and plants know *how to* engage in, while humans only have *knowledge of*, ecosystem engineering. It seems counterproductive to deforest an area that worked sustainably and was home for many organisms from fungi to mosses only supposedly to restore "dynamic" functions of all the geological elements (Gudorf & Huchingson 2003, p. 168). Sally Hawkins, et al. (2023, pp. 3–10) outline a definition and principles of rewilding that highlight the restoration of natural ecosystems. The definition presupposes a need to leave alone vast forest cores connected to ecosystems like grasslands and wetlands. These authors admit that any large forest ecosystem renovation is a "visionary" paradigm swing for radical change. Let's advocate for shifts in cultural attitudes that would obviate rewilding in the first place. A human-centered amputate-and-then-suture model is not nature at its optimal level. It seems maladaptive to cripple self-sustaining forest dynamos of global earth-enhancing air and water filtration and then labor to resuscitate those moribund areas.

Considering boundaries, rewilding could be part of reforestation for already damaged areas. Abandoned settlements can represent the "wisdom of environments in flux," says Cal Flyn (2022, p. 5), once untouched by human hands. We shouldn't pollute because we think we have technology to deflect the sun's rays or to vacuum carbon from the air (Davenport 2023). Carbon can be removed from the process of manufacturing cement (Reed 2024); but in the bigger picture, why produce extra cement to build more highways in newly deforested areas. Considering the true global costs/benefits, such plastic surgery is unnecessary if we shift attention to reengineering cities as sustainable centers, plot out edged forest areas for some human use with animal safety, and promote core forest growth and expansion for climate stability. John Reid and Thomas Lovejoy (2022, p. 4) define a core of the remaining five mega forests at 125 thousand square acres without any human constructions. On a smaller scale, in a process dubbed "succession," over time barren fields can cycle through softwood and then hardwood trees to become forests (Flyn 2022, p. 58).

In Estonia, notes Flyn (2022), after the Soviet occupation and withdrawal, there has been a gradual increase in forest growth on arable acreage that was previously community land. Likewise, this uptick in forest growth has occurred in Russia for similar reasons, but the important point is that those new forests of Northeastern Europe or Eurasia should be protected and preserved through sovereignty before a new wave of development destroys them. Caroline Fraser (2009), in a similar way, talks about the European Green Belt, wide strips of land from the Black Sea to the Barents Sea that were used for the Iron Curtain and have been restored to promote ecological diversity. More evidence that areas left to nature can rebound is seen in the demilitarized zone between North and South Korea, which over the years has dramatically flourished with biodiversity.

Politicians and corporate leaders typically work on short-term goals looking only for economic benefits, not spending enough money on sustainable development. There are exceptions, and the fossil fuel industry is beginning to realize investments are needed in green companies, just as the big tobacco companies began diversifying late in the twentieth century. Sustainable development is a long-term investment in global health, but the change might happen quickly with positive outcomes, as authors like Rockström and Klum (2015) suggest. Forest lands, which could include wetlands and grasslands, are assets worth preserving because of the healthy, climate feedback circuits they generate in terms of clean air and water. Large equatorial forests are water pumps to the atmosphere that stave off droughts; we see with our own eyes what has resulted from deforesting them.

Forest Crisis

After the ravages of World War II, Max Horkheimer and Theodor Adorno (2002) bemoaned how culture can regress and how the glories of the Enlightenment that began in the eighteenth century had increasingly caused people to become less self-aware. In particular, these thinkers warned that in the rapidity of scientific advance any meaning in life had been lost. Humankind had become godlike pushing its own form of sovereignty over existence where their power had come to estrange them from the world in which they live, what they term "social coercion" (2002, p. 9). The individual had been negated by the collective into a formula of logic, they feared. One of their main complaints is that reason had become the only instrument of functional purpose, evident primarily in machines of control. According to them, progress can lead to regression, where "civilization" ruins the "primeval world" (2002, p. 61). The self is diminished to material existence. Some of this thinking is particularly relevant since forest flora, fauna, and fungi have been reduced to commodities ravaged, bought, and sold by modern, industrial societies. In some cases, local communities, influenced by profit and materialism, are also leading to the downfall of the forests on which they should rely and respect.

More currently, philosophers Alice Crary and Lori Gruen (2022) observe that distorted views about how to treat humans arise from our mistreatment of animals. Social justice and liberation for animals are connected, and as one example they turn to Indonesia, noting how with modern machines forests are being cut and burned away to establish palm oil or sugarcane plantations. Farming other vegetable oils is also deleterious. This is a crisis of great proportions for the native species, biodiversity, and local people living on forest edges. Palm oil, which appears in products from food to shampoo by a number of different names, relies on deforestation, creating a threat to life forms near and far. Borneo and Sumatra supply most of the world's palm oil, and old-growth forests, which are home ranges for critically endangered orangutans and other species, are disappearing at alarming rates. Fossil researchers (Bochaton, et al. 2021) conclude that while Indigenous people have "coexisted" with wildlife for thousands of years without disrupting biodiversity, there has been "massive extinction" because of development and agriculture through colonization. In addition to the modern systems of forest degradation, the Intergovernmental Panel on Climate Change (Pörtner, et al., 2022) cites colonialism as a cause. As Crary and Gruen suggest, there's a long history of exploitation going back to early traders in gold and wood. Then, oil colonialism and capitalism enabled by state politicians fomented an enormous gap between the social lives of locals and corporate executives. Most people of Indonesia have not benefited from the explosive forest devastation and corporate growth. There has been no land or profit sharing, so there's a metaphorical pitchfork that skewers wildlife and poor people. As Crary and Gruen suggest, the plight of orangutans and other forest species cannot be separated from the nearby local people whose land and labor rights have been violated. The corporate pattern of mechanized capital over people and animals has been unfurling in other delicate forests worldwide, from Africa and Asia to Europe and the Americas.

Through education, aided by the musical, narrative, and visual arts, young people can understand that animal lives are not abstract and separate from human lives. Both are intimately connected with the same evolutionary ancestry. Animal ethics, say Crary and Gruen (2022), should not remain in academic discourse but broadcast loudly in a social sphere and connected to the climate crisis. In addition to palm oil plantations, forests are eliminated in other places to grow food for livestock (not people) and to house cattle slaughtered as human food. The Union of Concerned Scientists notes that twice the size of New Jersey in the U.S. is used just to grow chicken feed, and each year nine billion chickens are "produced" on yet more land creating various pollutions (Coulter 2023, p. 200). As omnivores, people do not absolutely need meat to survive, or they should at least dramatically reduce their meat intake to diminish deforestation (Machovina, et al. 2015; Tague 2022). People have been culturally programmed to consume lots of products that contain palm oil and to eat lots of meat. Consider, for example, the many advertisements over the years glorifying meat eating. Animals are sapient and sentient with self-awareness and agency (Jamieson 2018; Andrews & Monsó 2021). Breeding, raising, and killing them as food is socially unethical (Crary & Gruen 2022) since animal farming is one of the leading causes of climate change.

This leads one to question whether or not forests have moral status. For a consequentialist, value is based on the best outcomes. For a utilitarian, value is placed on function or pleasure over pain. For some thinkers, destruction of forests and wildlife lends them moral status. While forest animals have self-interests, they are dissimilar from those of humans, and this difference is seized upon as a reason to disregard them and their habitats. Speciesism, of which a utilitarian can be guilty, dispassionately sees the pain and pleasure of nonhuman animals, an unequal proportioning of human and then animal worth. What matters for the utilitarian and perhaps even a consequentialist is the overall result and not the individual, a form of moral irrelevancy that establishes systemic hierarchies between forest wildlife and human needs. Crary and Gruen (2022), from whom some of this line of thinking derives, suggest there are ideals like fairness, but we need to know under which circumstances to judge and apply equity. The time has come to give pride of place to forests because of the life support functions they provide for all organisms, whether in the forest or not. Just as animals don't have to be used as lab subjects or human food, forests don't have to be treated as corporate commodities. Normative considerations are not confined to a philosophy class but must be applied to Gaia.

The unfortunate truth is that forest-human interactions are governed by the politics of capitalism undermining the poor, minorities, and women. As insurmountable as it seems, these political structures of exploitation and destruction must be overhauled to recognize forest sovereignty, which includes the rights of marginalized and Indigenous people. As political philosopher A. John Simmons (2008, p. 44) says, sovereignty is "independence from outside control." Of course, he's thinking only in the human realm. If the print and social media would more frequently forefront environmental issues, there could be a call to action for change from people in democratic societies. Earth's destruction by humans is not coming; it's here. A moral revolution with social resistance against ecologically ignorant politicians must be waged so we can expand and preserve forests and create interspecies commons on forest edges, modeled according to sanctuaries and polyculture farms. While there are socially sensible ideas in the upcoming discussion of political philosophers, the overall aim is to develop a working theory and applied philosophy regarding the flora, fauna, and fungi in what's reckoned as the global forest state of Gaia. This posture does not represent a single political theory or ideology from the human heritage. Rather, evincing biological life's common origins and yet diverse evolution, there's a blend of autonomy and governance according to the dictates of living in proximity with other species. More protected forest areas help fight and stabilize climate change, as scientists report (Duncanson, et al. 2023); but without a political theory for the sovereignty of Gaia, forest management proposals are incomplete.

We have theories of justice and freedom in John Rawls (1999), Robert Nozick (1974), and Michael Walzer (1983) but nothing definitive for forest animals. In defining human nature, Nozick sees absolute property rights, with humans as solitary and deserving individual rights. One is entitled to all the fruits of her labor. In contrast, for Rawls, because people are born with different talents and abilities, some who work as hard as others do not equally gain; some people are born with gifts they don't really work to earn. Rawls looks at cooperation because humans are social beings with equal rights triggering reciprocity. For Walzer, principles of justice are pluralistic and dependent on the value and meanings placed on social goods. None of these perspectives is absolutely correct. There's a combination of these standpoints depending on circumstances. I try to align contrasting views of liberalism, libertarianism, and pluralism to demonstrate a theory of justice for forest organisms. My theory somehow blends the three in a slanted format. Nozick is right about forest animals: They've evolved adaptations that have entitled them to the earnings of their labors, the fruits of the lands they inhabit. Likewise, according to Rawls, humans foist an unequal advantage to themselves over the cooperative talents of forest flora and fauna that have evolved the immense state of Gaia, a self-regulating system of organisms. Acknowledging Walzer, we see how forest social goods are differently but deliberately distributed. Without the depth of political and ethical philosophy mentioned here, but somewhat similarly, Karen Bradshaw (2020) questions the assumptions that only humans can own land; worse, she asserts, lands owned by people kill biodiversity and drive climate change. From a legal standpoint and opposed to complete sovereignty, Bradshaw advocates for wildlife as property owners through a trust with executors working with judges to minimize abuses to nature.

Overview of Political Considerations

As in many modern societies, we see in a forest the liberalism of individual autonomy and free choice combined with republican ideals of a common, shared set of core values. Animals prize their liberty, but many species are quite group oriented. Liberty implies individual freedom and hence non-cooperation. Many modern individuals who value freedom nonetheless rely on outsiders to help them with decisions regarding mental, physical, or financial health. We can see the individual-group dynamic played out in forests, too, where the net result has always been positive for those who live there and overall global climate stability. While large countries might wield their absolute power over others, the system of Gaia is all powerful, even in its decline as evidenced in recurring climate catastrophes. The problem is that a majority of political philosophers up until the twenty-first century have mostly been focused only on the human, whether liberal, conservative, utilitarian, minority, Marxist, feminist, etc. Considering the age of this planet, the truth is that plants and animals are the rulers, not humans. This truth is enough to grant sovereignty to the sustainable forest state of Gaia.

In many ways, like the self-regulating systems of nature dubbed Gaia (Lovelock 1979), forest dwellers are a democratic collective where parts of connected ecosystems and ecological niches are advantageously self-governing. Vegetation influences and is not merely a byproduct of climate (Lovejoy & Nobre 2018). Deforestation has seriously upset hydrological cycles that generate rainfall, recycle air masses, and help cool the atmosphere. In some ways there's no real democratic coherence to the forest state, but biomes of forests work collectively for the benefit of the biosphere. Climate science demonstrates that without tropical forests and rainforests the earth's air and oceans are warming at disturbing rates (Nepstad, et al. 2008). There's no real oppression in the forest because, paradoxically, the self-interest of organisms helps others survive in cycles. Linear thinking industrial humans don't act that way and are often self-serving at the expense of others. Hunter-gatherers in forests tread lightly with no hungry machines chopping up lands and take only what's absolutely necessary, leaving room for growth.

Political philosophy might say, following Rawls (Wolff 2006, p. 164), that those entering a contract must be privy to the terms. There is no real contract between humans and the forest. One state (human) invades another (forest) with only justification fabricated according to perceived human needs, not counting Indigenous people who can work the forest equitably. To enlarge on thinking by Mary Midgley (2005, p. 369), rather than contractual responsibility for or libertarian freedom of individuals, we could shift our interpretation of life to a duty for earthly wholeness. There's no contract, no politics, no socialism, no egoism among wildlife. The forest, unlike human societies, is not a collective with legal and social contracts for individuals but a system of mutual interdependence of species, groups, and individuals, not profit-driven economies. The problem with maximization is that it involves a chance of loss for the risk taker, which is the scenario we now have with deforestation and climate change. In political philosophy and any theory of justice, it's fair to say that a human nation is not the same as the forest state since animal wants and needs are for survival, reproduction, mutual tolerance to avoid injury, and a certain level of care and comfort for self and others. This acceptance of others is not a low threshold but sustainable, in contrast to the excessive wants of many industrialized people.

Forest animals seem alike and willing to ensure liberty for all and equal access to resources given a large domain. In part, this is through evolved adaptations stemming from Darwinian selection and Malthusian economics so that there's no privileged class riding over the rights or access to resources of others. An entire tree, from the crown to its roots provides a home and nourishment for a wide variety of fungi, lichen, mosses, insects, bacteria, birds, and mammals. There is liberty and no strict, individual property ownership. Rather than respecting this symbiosis we have mostly ravaged it. The pattern of liberty in the forest is only "liable to disruption" (Wolff 2006, p. 172) by human consumption and intervention. The state of Gaia is not naturally disrupted. In the human state, Robert Nozick (1974) might say that free economic liberty and welfare distribution are not compatible (Wolff 2006, p. 174). John Rawls (1999) might say that freedom does not equate with anarchy. Walzer (1983) might say that the distribution of goods could be just across social structures but is often determined by membership. In the wild, welfare for others is not a constraint but an evolved adaptation.

Animal and plant forest dwellers have not consented to control; the human hand does not benefit the forest by pleasure or protection. Much animal activity is for any number of welfare advantages and distributions, like eating, playing, grooming, socializing, etc. There's partiality for valuable actions and experiences, Kymlicka observes (2002, p. 15). Humans violate these fruitful animal preferences. Helping people raid forests is of no use to creatures who live in symbiotic or mutualistic cycles with others. As just one example, weaver birds in Africa construct adjacent nests, and all can help maintain a common roof. It might be possible to say all forest agents act for others while satisfying their own needs. We see this because consequences of individual preferences have an overall benefit to the state of Gaia, which consists of a plurality of autonomous but connected species creating and maintaining a self-regulating biome.

The environmental stability of the forest is what controls all organisms in equal consideration of interests. Some may object by noting that my argument seems to favor the good health of a forest over needs of the individuals, but both are inseparable. Gaia provides many resources across species equally in open spaces since their evolved adaptations allow them to require different foods or parts thereof at different times. Generally, humans require and compete for exactly the same resources simultaneously to stockpile in private places and sell. In a large forest, no one is really harmed, but inequities of food and wealth distribution create income divides among humans. Usually, humans are not expected to finance the projects of other people, but that's what happens in a forest economy where the work of one, like an orangutan in a tree as a seed disperser, is the yield of another, like smaller creatures or microorganisms on the ground feeding from the ape's leftovers or fecal droppings. There's no infringement on the preferences of others, unlike similar species who feed from the same resources. Even so, a chimpanzee might eat the fruit of a plant while another, like a lowland gorilla, eats the leaves. Assuming the forest has not been reduced by development, each has a fair share, a political goal with some moral consequences for all (Kymlicka 2002, p. 43). In a human society, some people will deprive others of resources; or, because they lack skills, some humans simply cannot achieve the resource gains of others. This is not so in the forest except when humans intervene to deprive animals of much-needed land for the resources to flourish. Overall, there is equal treatment in the forest without the burden of humans.

Building a highway in the forest to transport consumer goods like timber not only damages the ecosystem but also presents a physical barrier and hazard to animals. No forest organism should be bound by the human state, only its own. It's not as if animals can migrate to another forest given the magnitude of fragmentation imposed upon them by humans. People in modern nations have benefited from the resources and materials in forests; human incursion has not been a boon to animals. Other than populations of Indigenous people or hunter-gatherers, the citizens of the forest who claim residence are the diverse flora, fauna, and fungi, so it's their state. In *Politics*, Aristotle says good citizens engage in participatory commitments for a "common enterprise" (S. Smith 2012, p. 78), precisely the merging of public and private interests in nature. In his *Ethics*, Aristotle also implies how the common good, without discounting individual well-being, is a higher perfection, the state over the individual, as Plato might say (Winton & Garnsey 1981, p. 50). Individuals matter insofar as each contributes to the functioning of a state that fares well. Animals should not be considered as refugees in the forest, nor should the forest be considered their refuge. According to political philosophy, the forest is rightfully their state. They've inhabited the place for many millennia and have been the ecosystem engineers freely maintaining its natural resources.

Which nation-states will admit all the forest animals with open arms and grant them safe refuge were all forests eradicated? While many states take in human refugees from nations at war or offer political asylum, the human heart is not as kind to wild animals. Of course, there are many exceptions from the individual to organizational level, predominantly in sanctuaries, but their existence begs the question of forest sovereignty. Thus, forest animals and all the other inhabitants, whether microorganisms or old-growth trees, should be granted dominion. The forest is a state for animals because, all things considered, it offers freedom, peace, and security, triggering adapted emotions and feelings. While there might be species and individual differences, science shows that animals experience subjective feelings in the wider context of bioregulatory adaptive emotions (Kret, et al. 2022). In modern states people are provided with freedom, peace, and security. If humans have earned a state of sovereignty, so have forest animals.

In one sense, forest animals do not have to struggle for freedom through political debates and litigation as do many people in developed countries. Industrial nations falsely perceive the flora, fauna, and fungi of the forest as without any government. Indeed, there is none similar to what's seen in human nation-states, but certainly there is governance in ecosystems under the umbrella of Gaia. We just can't see or ignore that governing system. Similarly, the first European settlers in the Americas intentionally and with clear understanding aimed to destroy the lives, cultures, and food sources of Indigenous people; they spread disease and viruses and cared not to understand the forest cultures or environmental health of native inhabitants. Is it any wonder that when a palm oil or sugarcane plantation company razes a forest, primates who have lost habitats and food raid human crops. Yet they are considered pests. Animal rights and needs are not considered during, nor are they included in, agricultural plans as deforestation takes place. Bears, deer, opossums, beavers, racoons, etc., face similar fates as suburban sprawl into rural areas displaces them.

As in generations past with children, women, and minorities, forest inhabitants have been politically disenfranchised. It won't be sufficient to have a political theory of forest sovereignty without results. Instances of corporate or agricultural deforestation deny all wildlife full participation in their homes and ways of life. There is little to no due process of law for forest dwellers, whether flora or fauna. With their essential freedoms circumscribed by habitat loss, forest inhabitants cannot govern or act for themselves as they evolved to do in

Gaia. Like many political states, the forest offers its inhabitants a life of well-being. That's not hypothetical or perhaps not quite ideal but real, and animals exert free agency to stay in the forest voluntarily unless pushed out. It's theirs, a decentralized state with multiple mutually dependent structures as evolved adaptations of influence and control.

Carol Gigliotti (2022) offers the example of sanctuary horses and chickens. The birds aid the equines in parasite and pest control while the horses protect chickens from predators. Parasites are part of an evolved biodiverse ecology by regulating host overpopulation, encouraging energy synergies, and promoting food system connections (Wood, et al. 2023). Human meddling with landscapes is disrupting how species interact, confining them to less space, instigating unanticipated antagonistic interactions, and elevating stressor consequences (Gilbert, et al. 2022). Although there's no forest state defined by human terms and boundaries, wild animals nevertheless act as if there is one. In spite of what political philosophy might say, there was never chaos or anarchy in forests when they were large enough with plenty for all. If there's chaos, it appears in self-styled civilized societies. One news article (Einhorn 2024) bluntly states: "Animals need humans to solve the climate crisis." My argument is the opposite. The forest state of Gaia is a key contributor to restoring climate health, not necessarily human technology.

Legal Issues Underlying Sovereignty

In *Wildlife as Property Owners*, legal scholar Karen Bradshaw (2020) lays groundwork for the claim of forest sovereignty, but there are qualifications. While Bradshaw's book touches on biology and political theory, it's closer to establishing a legal footing for animals. My focus is on the forest systems of Gaia that include not only animals but plants, trees, fungi, moss, lichen, microbes, insects, etc. Bradshaw addresses these elements briefly in her Conclusion, so my argument picks up where she leaves off. A key difference is that Bradshaw is arguing for something—legal rights—that should be apparent already in nature since Gaia has functioned with modified property interests (e.g., ecosystems, niches, etc.) for millennia. She argues that the best remedy is to create wildlife habitats, but these would diminish my argument as smaller versions of more and larger core forest areas. In other words, Bradshaw argues for laws that allow animals to own land with a trustee, but there's no political theory to bolster that position. Bradshaw's argument is persuasive and laudable, and what's good is how with her plan land could be transferred to an animal trust.

Yet, it's difficult to grasp Bradshaw's (2020) views as a final solution. Using her line of thought, metaphorically, common law permits property transfer through inheritance; per Darwin's (2003) observation of variation, selection, and heredity, earth has been transferred from one generation of plants, animals, and microbes to another continuously. In multilevel selection, benefits fall to the group regardless of the effects on individuals; that's what has been happening in Gaia over eons. A key question for legal scholars is where property rights originate and when they are relinquished. They began in nature and are not surrendered there, as Bradshaw agrees in what she calls (without elucidation and focusing more on territory marking) the biological paradigm of property rights. Looking at the whole effect, flora, fauna, and fungi of Gaia's biosphere have evolved ecosystem properties and biomes working in tandem with equanimity. Large dominions create territories, so by implication, sovereignty includes rights. Somewhat contrary to zoopolis or animal sovereignty (Donaldson & Kymlicka 2011), forest sovereignty is the final solution.

Forest sovereignty, unless actually codified, is not currently by statute but by usage. William Blackstone (1979, p. 2) says that property consists of "prolonged occupancy and investment of labor." Unfortunately for sovereign Gaia, the human notions of land use and entitlement have led to deforestation. We don't need welfare for nature but require property rights, as Bradshaw (2020) strongly suggests. Bradshaw's lens is somewhat foggy, though, since her notion of property rights in the biological sense weighs too much on territoriality (e.g., defense, collectives, housing, etc.). Her argument for property per se would of course rest on boundaries, and that's the problem since the human perspective establishes property lines to demarcate one's possessions from another's. This legalistic and nearly anthropocentric outlook ignores the mutual aid and sharing behaviors in nature; with plants and animals it's not constant bloody competition, especially not in how humans compete to expand land holdings.

In going further than arguing for animal property rights, look how Gaian forests, not to discount oceans, are necessary to sustain fresh air and clean water for all life forms. While ecosystem levels are important to consider, the logical conclusion is to argue for forest sovereignty ensuring the continuation of all biodiversity. We are stakeholders in this fight and life itself is at risk. Some redistribution of land is inevitable with the forest sovereignty proposal simply because humans have proved to be poor managers of nature and need to acknowledge their shortcomings. Contrary to Adam Smith (1952), property does not always have to be part of a humanly manufactured government; if it is, let the land pursuant to natural rights become immense sanctuary habitats. Bradshaw (2020) is correct to claim property rights for wildlife, but that's just one step toward the necessary, ultimate outcome of forest sovereignty.

In the next chapter, let's examine what's central to forests: plant life. Then, in subsequent chapters, we'll see how political philosophy has not considered preserving forest vitality. Recalling the proposal set forth in the Preface, in many parts of the world Indigenous perspectives, practices, and ideas that foster nature's health need to be included in discussions concerning the natural world. Political theorists are summoned to develop a line of argument about forest governance; that is, there's an environmental critique of Enlightenment and social contract theory to advance a non-interference approach for the forests of Gaia. The attempt is to read political philosophy against the grain by exploring anthropocentric concepts and offer ecological theory using the very ideas from these seminal thinkers.

Chapter 1

The Forest State of Gaia

Great forests are transnational and should be conserved at all costs. At the same time, smaller areas need preservation, too, since they serve as homes to a multitude of flora, fauna, and fungi. Leaving nature intact and undeveloped without human manipulation will permit biodiversity to run its ecologically beneficial adaptive course (Anderson & Ferree 2010). In the forest collection we call the Amazon, during April there's so much atmospheric and land hydration there are flying rivers (Salati, et al. 1979). Reforestation, though encouraged along with rewilding, does not ensure biodiversity restoration; it's more feasible ecologically and economically to protect and expand existing great forests (Reid & Lovejoy 2022, pp. 237, 246). In addition to tropical rainforests, like those in the Amazon and Indonesia, others are located in places like the Great Bear Rainforest, northern coast of B.C.; Central America; Africa; Europe; Russia; China; Fishlake National Forest, Utah; Ala-Archa, Kyrgyzstan. This chapter demonstrates the self-regulation of inherently governed forests and is an important precursor to the chapters on human-made political philosophy that has disregarded but can be used as an argument for the necessity of wildlife sovereignty. Nature should not be relegated to the needs and desires of modern people, as we now understand.

Ecosystem Engineering

Ecosystem engineering refers to how an organism creates and modifies its habitat or modulates the supply of resources (Jones, et al. 1994). Beavers are well-known eco-engineers who ably maintain the stability of a locale, which consequently affects the distribution of resources to other species. Ecology is the interdependence of organisms. Peter Wohlleben (2019, p. 183) refers to thousands of forest microorganisms, bacteria, mites, lice, fungi, and insects that help decompose fallen trees, leaves, dead animals, etc., into humus as "sanitary engineers." Darwin was aware that earthworm activity builds soil formation. Worms, says Darwin (1989), demonstrate what seems to be intelligence above instinct in how they handle foreign objects with purpose to bring to their burrows. Though blind and deaf, worms have the ability to experiment with objects to their advantage, a sign of intelligence. In burrowing, worms obtain nutrients by eating the soil. They also consume almost any matter in order to burrow.

Quoting other sources, Darwin (1989) makes some calculations. There could be over 53,000 worms per acre, at least in gardens, with about half that number in a field. In a year, approximately 7 to 18 tons of earth per acre are cast upward from the work of worms, depending on the time of year and other conditions. Darwin says these creatures play an important role in nature, for they aerate soil, germinate seeds, allow moisture to penetrate tree roots by tillage, and help mingle and distribute nutrients, chemicals, and key microscopic organisms necessary for plant growth. On a larger scale, Joshua Barry, et al. (2018) show how pumas kill large prey, eat what they need, and then leave the carcass as food for hundreds of species breaking down remnants of flesh, bone, and fur for soil nutrients. All organisms in a forest, from bacteria to mammals and large ungulates, sustainably manage the trees, plants, soil, etc. Eco-engineering does not only include some "feedback effects" to the engineer but also net results for the locale and other species, especially in a large area, what Clive Jones, et al. (1997) call the "flows of energy."

Let's look at the opposite of ecosystem engineering by examining what Brian Machovina, et al. (2015) say about animal farming by humans. Habitat, ecosystem, and biodiversity loss are primarily driven by livestock production evident from data covering 1700 to 1990. Livestock farming generates dangerous amounts of anthropogenic greenhouse gases, a little over 14 percent. This unnatural change to the climate is compounded by land use degradation with subsequent contaminations from fertilizers and other farming chemicals, causing poisonous runoff exacerbated by flooding from global warming. Natural grasslands and forests have been greatly diminished as land was increasingly converted to food production for humans and then to a greater degree for animals to be slaughtered in factories as human food. Grasslands cover much of earth's non-ice surface, about 40 percent, and store carbon, but these lands are regularly used for human but mostly livestock agriculture releasing harmful carbon dioxide. Simply increasing grasslands for soil carbon capture is not enough since methane and nitrous oxide from ruminant livestock feeding and grazing offset benefits (Wang, et al. 2023). A solution is to reduce cattle farming as a food supply so in their place expanded grasslands can mitigate greenhouse gases.

Likewise, terrestrial forests and water ecosystems capture 6 billion tons of carbon each year, but human activity pushes 11 billion tons of carbon into the air; this disproportion created and worsens the climate crisis (Gonzalez 2023a). One report (Jay, et al. 2023, p. 42) indicates that climate change problems can be solved by reforestation of nature-based carbon storage in woodlands and wetlands. By implication, therefore, forest flora and wildlife fundamentally constitute ecological sustainability, not humans. Research scientists unequivocally demonstrate that restoration and conservation of forests and soils present immense potential for increased storage of carbon on land (Walker, et al. 2022; Mo, et al. 2023). Preservation of existing lands is more crucial than simply planting more trees. As one might expect, from the science side there is scant analysis about political, animal, or ecological ethics, as offered here.

Citing work from Russian, European, and American scientists, Ben Rawlence (2022, p. 94) notes that plant life can be overwhelmed by too much carbon. At the same time, much precipitation, a product of global warming along with drought, will stress tree root systems, as we see happening in real time from weather reports. The Union of Concerned Scientists (Boucher, et al. 2012) admit that while up to 20 percent of earth's agricultural land produces human food, four times that amount of land is used to feed livestock fated for slaughter as human food, to say nothing of massive tracts of land deforested for and trampled by cattle, reducing vegetation and increasing erosion. These concerned scientists say that meat production is "ecologically inefficient" but nonetheless advocate for animal agriculture of pigs and chickens; so, ethics is not their concern. According to Our World in Data (Ritchie 2022), over 60 percent of global biomass is livestock consuming disproportionate amounts of grains and water. Wild mammals who are integral components of Gaia's eco-engineering constitute only 4 percent of the earth's biomass. The food culture of animal agriculture with its massive deforestation and waste is not an ecologically sound system. With little end in sight, more than 75 percent of deforested Amazon land is for livestock grazing, or crops to support them as meat demand escalates. There's also water waste in animal farming. In Africa, tropical deforestation is driven more by logging, but this could change and follow models in Brazil for livestock. There's also a large bushmeat trade driving forest biodiversity loss in Africa and southeastern Asia. Like animal agriculture, this hunting destabilizes ecosystems (Darimont, et al. 2023). As far back as the 1830s, Robert FitzRoy and Darwin saw how in South America wild pigs and goats hunted by colonists were becoming decimated in unsustainable populations (Preston 2022, p. 262).

All these factors lead to a breakdown of soils, water depletion, and pollutions to air and waterways. Human disturbance of forests impacts soil feeding levels below and above ground with bad consequences for a multi-trophic ecosystem (Laigle, et al. 2021). The bottom line is that deforestation on such a massive scale has disrupted, perhaps permanently in some places, the natural state of Gaia. Deforestation equates to carbon emissions and "ecosystem stress," all of which if left unchecked cause permanent damage to world Gaia (Gatti, et al. 2021).

Peter Kropotkin, *Mutual Aid* (1902)

Peter Kropotkin (1902) believes a key factor of evolution is mutual aid. Too much emphasis is placed, he says, on distorted readings of what Darwin calls the "struggle for existence," not physical combat but a metaphor of adaptation. Kropotkin admits, like Darwin, that an intense fight for survival entails suffering and death. No one disputes that reality. However, competition alone does not fully explain aspects of species survival, reproductive success, and adaptation. By the end of his Chapter 3 in *Origin of Species*, Darwin (2003) suggests that despite struggle there's more of what he calls a uniform proportioning of resources to organisms. As Ernst Mayr (1997, p. 199) says, through competition organisms evolve survival "compromises." Darwin goes on to say how all organisms in nature are meshed together in a web of communally distributed forces. The overlooked aspect of evolution Kropotkin underlines is mutual aid, of which there are countless examples in nature. For my argument, this support network, sometimes among different species, is what enables the ecological engineering of Gaia to function as a self-regulatory system supporting planetary health. Survival can be a mutual enterprise.

From his Chapter 1, the examples Kropotkin offers are many, so let's mention only a few: combined efforts in feeding or nesting, or feeding strangers; groups forming a ring to stave off an attack; herding, schooling, breeding, and migratory associations; multiplied forces; social instincts and socialization; rearing young; watchfulness; attachments; safety controls in solidarity; orphan adoption. These behaviors and more run across and sometimes between species, from insects to birds and mammals. Darwin noted many examples of mutual adaptations between nectar-providing plants and cross-pollinating bees (Costa & Angell 2023). Furthermore, Kropotkin insists that these activities evince, to borrow his words, justice and equity where one acknowledges another's rights (1902, p. xiv). Two points are worth noting going forward. 1. Modern humans have upset any balance of mutual regulation in nature's ecology that guarantees earth's health. 2. Along with number 1, Kropotkin's field observations of equitable behavior in nature get to the heart of any argument for forest sovereignty.

James Lovelock, the Gaia Hypothesis (1979)

By 1979 earth scientist James Lovelock (2000, p. vii) working with microbiologist Lynn Margulis concluded that as a "consequence" of earth's living diversity there's a self-regulating system of comfort to enhance fitness for habitable conditions. This is still a radical idea that aligns with Thomas Kuhn's (2012) notion of a paradigm shift in scientific revolutions. In his 2000 Preface to *Gaia*, Lovelock says that consumer demands drive corporate growth for forest products; he quickly adds that with industrial production there is little regard for the environment. With proper education and cultural modifications, we can reverse this trend. Gaia is alive as a gene is self-interested in its own survival in unconscious operations, declares Lovelock. He says the first edition focused too much on living inhabitants and was erroneously neglectful about sources of climate control like water, ice, clouds, seaweed, rocks, etc. Here, for the sake of simplicity, I focus only on forests, assuming those areas include wetlands and grasslands. At several points in his Preface, Lovelock refers to Gaia as a superorganism in its animating energy and mutability. E. O. Wilson (2016, p. 13) refers to Gaia as "density-dependent regulation" where all life populations interact. If, according to the laws of natural selection, advantageous variations are passed on, then by analogy the biosphere is a superorganism that has been, through its billions of living beings, selected to thrive. Let's see the forest as a superorganism entitled to ethical consideration.

In Chapter 1, Lovelock tries to answer the question, what is life. His tentative answer goes something like this: a process located in and formed by the "abundant flow of energy" (Lovelock 1979, p. 4). This does not include machines but recognizes weather events, transitions in soils, rock formations, as well as movements of organic life forms. He notes that in 1966 Shell Research Ltd. commissioned him to opine on air contamination from the burning of fossil fuels, since corporations already knew about pollutions. Lovelock's hypothesis is that all of earth's matter functions to manipulate the atmosphere (technically, the troposphere) adaptively in a type of bio-feedback sustaining itself. Lovelock says the earth's atmosphere is likely not some random occurrence but generated and sustained by Gaia (1979, p. 62). Atmospheric optimum is not by chance alone. The earth's climate including overall and regular temperature and atmospheric conditions has been sustaining life for 3,000 million years (Lovelock, 1979, p. 18).

There's no pollution, as we'd categorize it, in nature. The dung of one animal is food for another with "chemical neutrality" (Lovelock 1979, p. 26). We've offset that balance. Lovelock goes on to say that flora and fauna are far less important than what lies beneath and within the soils and continental shelves since those organisms are the prime movers of life (1979, p. 37). The focus here is on forests since in their remaining vastness and presumed expansion through protection we can restore planetary health. As we will see, fungi in soil composition are critical to the health of trees and in consequence to forests. By concentrating on forests, we can direct attention to everything there, from soil microbes and animals to plants and trees. There is no singular part of a forest that is important. The point is that what we see and call a forest is an evolved web of interlocking and mutually beneficial activity on multiple levels. Harming one part can damage the whole. Lovelock (1979), who expresses optimism in his book, asserts that we will undoubtedly persist releasing industrial pollutants to destroy Gaia (1979, p. 42). He confirms how a multitude of species when attending to their daily routines cooperate to sustain each other as well as all forest matter (1979,

p. 40). Lovelock reminds us that the only organism with a dangerous appetite to kill Gaia is humankind.

Oxygen molecules arise not just in photosynthesis and water vapor in the atmosphere but also in the death and movement of organic plant matter (Lovelock 1979, pp. 65, 69, 72). Gaia uses but does not waste energies, chemicals, or gases of biological origins that help regulate climate and temperature. We know these natural cycles are harmed by the aggressive burning of dangerous industrial fuels increasing acidity in rain, which then reduces green growth. Pollution occurs naturally in gases, chemicals, or radiation. Early on, oxygen was a poison that forced microbes to the sea floor or into mud. These tiny organisms are now in animal guts. Lovelock wondered in 1979 if pollution is irrelevant to Gaia. Back then, he was optimistic about adaptations to some harmful effects. Vegetation bides time. But on scale, any organism, including Gaia, can withstand only so much industrial waste. If forests are cleared, the time it would take for such extensive regrowth, were it to happen, is immense, in addition to the repercussions on that area from so much biodiversity loss. In Lovelock's time, there were still questions about global warming and even suggestions that hazy pollution might cancel out warming. None of this is true. Microorganisms are crucial to Gaia and will persist with any rapid de-escalation of flora and fauna, but why should we do nothing to see if that prediction comes true.

Lovelock talks about potential concerns (1979, p. 108), but the attitude of delaying panic has contributed to our current crisis. He even suggests leaving issues of the climate to lawmakers (1979, p. 109), but we see the dire results of legislation, no matter how well intended. Backpedaling, the European Union passed legislation to restore 20 percent of its land and sea (EU Parliament News 2024). Political indifference is also a cause of environmental damage. A large part of the problem is not governmental agencies like the Environmental Protection Agency in the U.S. but the politicians and lobbyists representing the oil, gas, coal, mining, and animal agricultural industries who oppose environmental laws and sue at every turn. Bad politics by flawed politicians influenced by lobbyists focused on corporate profits are parts of misapplied science and culture. At the same time, in Chapter 7, Lovelock recognizes that core regions of forest Gaia have been compromised. He sees human danger in forests or the continental shelves, not in cities. Yet, urban areas produce industrial pollutants and high levels of carbon dioxide and are homes of massive consumption, leading to forest mining and logging. Nevertheless, green, sustainable cities are likely the wave of the future since population concentrations there can permit rewilding and reforestation elsewhere (Benyus 2022; Attenborough 2020, pp. 208–10).

In reference to cooperation and mutual aid in Gaia, Lovelock notes how organisms relate to one another in their environments (1979, p. 115). These are salutary interactions and not the failed notions of human stewardship of earth. Gaia is focused on microorganisms, but they are part and parcel of mega flora and fauna. Is there a human ecology separate from Gaia? That's hard to answer considering how human activity poses serious risks to planetary health. At the same time, were all humans to go extinct, Gaia would continue in some form. Although humans are part of Gaia, forests can thrive without humans. Lovelock says Gaia is bigger than ecological studies focusing on habitats and ecosystems, yet he wonders if humans have not "seriously interfered" with the homeostasis of Gaia (1979, p. 119), a clarion call that was ignored until only recently. Some might say no; some might offer a qualified answer; we have, in his words, created an "inertial drag" (p. 119) with malignant consequences.

In his Epilogue, Lovelock (1979) ponders how our intelligence is part of Gaia. This is a complex question to answer, but an attempt was made by E. O. Wilson in his book *Biophilia* (1984). We are a species like all others derived from earth's elements. Darwin recognizes plants as distant, common ancestors to mammals in their abilities to move, their sensations, and the carnivory of some, attesting to the diversity and variety of adaptations (Costa 2017, pp. 264, 324). Our *Homo* line survived over others but has not yet lived as long as our ancestral relative *H. erectus*, an early hunter-gatherer. We love natural beauty. We care for animals as pets, recognizing their mammalian proximity to us. Complicating this rosy picture is our consumption on an enormous scale, eating away and polluting our natural home.

Lynn Margulis, *Symbiotic Planet* (1998)

In light of Lovelock and Margulis, it's worth looking at *Gaian Systems* by Bruce Clarke (2020), a book that demonstrates how Gaia theory was hotly debated in the closing decades of the twentieth century. I will focus on parts pertinent to my discussion, but one theme from Clarke central to my undertaking is clear: Gaia is biopolitical. Following science philosopher Isabelle Stengers, Clarke aligns with some of my thinking by stating that while earth resists capitalism, how much stress can the planet bear. Like the net effects of ecosystem engineering, Gaia deals in quantities and results, so there is some but not a total loss of plant and animal life, says Margulis (Clarke 2020, p. 230). Rapport comes through the acceptance of differences, not enforced commonality. Humans, therefore, are not immune from having compromised Gaia.

Opposing Lovelock and Wilson, Margulis (1998, p. 105) does not see one organism in Gaia but a web of symbiotic ecosystem networks recycling "biologically important elements." For the sake of argument, let's liken this interconnected grid of ecosystems to a superorganism. At any rate, as a microbiologist Margulis was influential in helping complete Lovelock's thesis of the planet's atmosphere as dependent on living organisms like microbes. Along with Lovelock she posits earth's "thermodynamic equilibrium" (Sagan & Margulis 2023, p. 3) because of how micro and macro-organisms as well as plants use the atmosphere to survive. Gaia is a coordinated system of chemicals, matter, water, air, and life forms maintaining planetary thermoregulation. Dorion Sagan and Margulis (2023, p. 6) go as far as suggesting that earthly Gaia reveals a naturally purposeful "mind" evident in trees that bend toward light to control and disperse energy streams of heat and moisture. Even on a microbial level, organisms recycle and reformulate the environment of atmospheric gases to regulate and stabilize earth's temperature (Sagan & Margulis 2023, p. 26). Gaia is like a vital, metastable body containing systems of interacting bodies.

Gaia, according to Margulis, is like a collective body of systems, or what I'm calling a state of ecosystems. Clarke (2020, p. 4) emphasizes autopoiesis in Gaia. Living and non-living components of the system are altered by the structure they continuously reform in a self-creating process. For Clarke, like biologist philosopher Francisco Varela, Gaia is a system of interactions supporting life, autopoietic recursiveness emerging invention and persistence from rocks and soils to organisms like microbes and trees. Taking Lovelock and Margulis together, Gaia is less biocentric and more modular where various biotic components and abiotic reactions regulate the environment that in turn affects them. Margulis asserts this in her theory of how procaryotes like bacteria combine to form eukaryotes symbiotically. Key to Margulis is how bacteria do not always compete but cooperates in non-linear recombination (Clarke 2020, p. 160). Where Lovelock sees a macrocosm, Margulis sees the microcosm. Following this thought, Clarke (2020, p. 48) calls on Donna Haraway to claim that Gaia is mechanical and organic. Others disagree. Isabelle Stengers, Clarke goes on (2020, pp. 53, 57) does not see a cybernetic Gaia but a transcendent though not immanent "being." Gaia is alive, but yet not quite, perhaps why I refer to it as a state.

In his review and complicating this picture, Clarke (2020) discusses Bruno Latour who does not see Gaia as a totality but asks, point blank, about our politics regarding earth. Gaia is about metabolic-like connections, which for Latour does not necessarily mean one whole (Clarke 2020, p. 67). Contrary to Latour and how I grasp Lovelock and Margulis, there's a type of bio-mechanical order to Gaia, a correlation of inorganic and other systems like air, water, soil, flora, fauna, and fungi. Latour does not favor the appellation "superorganism." However, what we call the unbound parts are really the environmental whole through the functional systems of Lovelock or the symbiotic genesis of Margulis. Overall, the parts work together in self-interested participation, mutually creating feedback energies. If Gaia is autopoietic, it is not then a superorganism, says Clarke (2020, p. 93). For the sake of simplicity, some people refer to the enigma of Gaia as a superorganism though it's not a singular body per se but a complex arrangement of moving and changing parts, a network. At the same time, organisms, chemicals, and minerals are constitutive of and create earth and the environment. So, one can't disagree about the completeness and yet fragility of Gaia, as argued here.

Aldo Leopold, "The Land Ethic" (1949)

The use of nature's property has been dominated by a human sense of expediency with privileges and no obligations of right or wrong, Aldo Leopold (1949) says. Drawing from Kant (Kuehn 2001, p. 313), we might say nature favors moral autonomy since human morality is conditioned on so many expedient values, beliefs, and desires. There's an instinct to compete and an ethics to cooperate. Which of these postures predominates in the human or wildlife worlds? Leopold suggests that any community ethic should include the land. Indeed, following Leopold, a land ethic would guarantee the existence of the flora, fauna, and fungi and convert humans from forest conquerors to citizens of a shared ecological community. Leopold (1949, p. 208) is critical of human "visible economic gain" over ultimate ecological conscience. Corporate farmers, loggers, miners, etc., use land as property of profit over a land ethic, self-interest over long-term conservation. This negative ecological attitude comes about because most of the plants and animals on any given land cannot be sold or eaten, and so they are often destroyed to establish animal agriculture or harvests to feed factory animals, upsetting the stability of an ecosystem.

Leopold goes on to say that wild plants and animals of no apparent utility to humans nevertheless have a "biotic right" to exist (1949, p. 211). He notes that some trees in the U.S. during his time were of little worth to timber companies and not, as they would be in Europe, valued for their contribution to a native forest's soil and bioecology. We tend to consider the forest as a commercial reservoir to be emptied according to human desires. Likewise, grasslands and wetlands are presumed to be of little economic value but, on the contrary, make immense contributions to an area's biodiversity network. Leopold's argument is that nature in private hands could be valued for its own sake; indeed, we see hundreds of thousands of hectares being acquired for preservation in some places, like Borneo (TOP 2024). The state or federal governments should not be the sole caretaker of lands that contribute to biodiversity and help stave off climate change through carbon capture. Private owners, he suggests, have a duty to the natural and human communities supplemented by government involvement with financial support or tax incentives.

If, as stated by E. O. Wilson (2016), the forest is a superorganism, amputating parts of "value" for humans damages the whole. Leopold (1949, p. 215) says food chains are complex and seemingly "disorderly" but evolved as elaborate systems of both competition and cooperation. Nothing is wasted in nature. As Darwin (1989) implies in his book on worms, soil is in constant movement, a force of energy that supplies life and feeds earth's surface with food, water, and air circulation. All of the various life forms depend on each other in their lifecycles. Leopold (1949, p. 216) describes this circle as a fertile circuit where one small change affects the entire conduit of energy. While this change could occur naturally, organisms will make adaptive adjustments. If, on the other hand, there's a violent change, such as deforesting an ecosystem for human industry, the result is depletion, erosion, and extinction. Regarding the land ethic, Leopold reminds us that what we call land is not simply flat earth but an array of plants and animals feeding from and contributing to the fertile microbial energies of the ground. He admits, as has occurred in parts of Europe, that lands develop a new *modus vivendi* and survive. This conversion predates the mega-machines of the twenty-first century that completely bulldoze thousands of hectares in violent rapidity of disorganization where the land never recovers but merely becomes an instrument of political or capital gain.

Leopold (1949) laments how education and economics have turned away from a biotic consciousness focused on a land ethic because most modern people are physically separated from proximity to wild lands or live in manicured housing developments. The land problem is determined by human economic attitudes and their machines, but that view of earth can change through cultural evolution with a greater emphasis on the forest sovereignty of Gaia as a key component to climate sustainability. According to Gudorf and Huchingson (2003, p. 11), Leopold can be considered a consequentialist (i.e., what will happen if forests are destroyed) as well as a deontologist (i.e., promoting right action to preserve an ecosystem). Each organism functions for itself but in its ecological niche services the entire living network, from bacteria to bees. Leopold exemplifies how political philosophy made by humans works well for us, but for the forest we require a renewed theory of Gaia.

E. O. Wilson, *Biophilia* (1984) and *Half-Earth* (2016)

Drawing from the land ethic of Aldo Leopold (1949), biologist E. O. Wilson demonstrates human continuities with biological nature and the need to preserve most of the wild. Wilson says that what he terms as biophilia is the human predisposition to ground our physical and emotional experiences in the processes of life in a non-rational manner. We are intrigued by other organisms and begin to value them more as we grow to understand their complexities and connections to us in a shared biosphere. We still might not fully understand what Wilson refers to as the "energy network" (1984, p. 5) in forests near human habitations.

Indigenous people seem to comprehend; perhaps up to the early eighteenth century early modern humans did; but now the human-forest codependency seems lost. To avoid catastrophic climate changes, the remaining 35 percent or so of forest lands sustainably managed by Indigenous people should be left intact (Fa, et al. 2021). The forest is another world and yet not, considering how our australopithecine ancestral relatives partially emerged from the woodlands a few million years ago to investigate resources in the nearby savanna.

On the microbiological level, Wilson (1984) notes, there are millions of tiny soil organisms that live and die, assimilating themselves in the ecology of food for others in cyclic interactions. To mitigate competition in what Darwin (2003) calls the struggle for existence, forest species, from those in soils to tree canopies, have evolved adaptations of coexistence. As Kant might say (Kuehn 2001, p. 397), there's a concurrence of freedoms. More life in any one spot of earth emerges from diversity; overall, in this struggle of varieties, stability arises (Costa 2017, p. 109). Now, we mostly don't coexist with forests and their teeming array of life. Yannick Joy and Andreas De Block (2025) are critical of the biophilia hypothesis as an innate feeling for nature spread by natural selection. They incline more to preferences through cultural evolution. This postulation bolsters my argument since our mammalian propensity toward nature requires enhanced cultural learning for earth's verdant survival. Perhaps biophilia has run its course. In the 1980s when he writes, Wilson suggests that through culture we are lost between city machines and forest nature from which we evolved. There's an ever-increasing and dangerous tilt away from nature and toward technology. With our attention diverted to technological gadgets or other material goods, we've grown to ignore a sense of nuance in the smallest spot of earth. We rarely look at Gaia directly except through a miasma of projected images in paintings, movies, verbal narratives, advertisements, virtual reality, and videos. As Wilson sees it, humans are self-ignorant and stumble through their current cultural evolution emphasizing discord with or alienation from nature.

Regarding the notion of Gaia as a superorganism, Wilson (1984) bluntly says extinction accelerates as habitats decrease, and while decline can occur naturally, most of it now happens because of human interference. A superorganism, like an ant colony, has intelligent parts divided according to rank, size, and labor in a "balanced whole" (Wilson 1984, p. 36) where brain power is spread throughout multiple organisms. The forest consists of "species equilibrium" and biogeography or the filling of ecological niches by an array of life forms, Wilson notes (1984, p. 39). Drawing from Aldo Leopold (1949) and his land ethic, Wilson says that rules governing the treatment and regulation of earth must encompass the needs and desires of future generations, not the expediency of the present. Writing during the Cold War, Wilson perspicaciously says that the tragedy to come is not nuclear war but the destruction of species habitats. Current mass extinction is not Darwinian or Malthusian but driven by epidemic forest destruction by humans. As Wilson predicts, within our lifetime we're on track to destroy Gaia. A handful of politicians guide environmental policy that should be ground in science, ecological ethics, and perspectives of Indigenous people.

In *Half-Earth* (2016), Wilson places the crisis of biodiversity conservation at the center. In *Biophilia* (1984), he praised our connection to life but warned of threats to Gaia, whereas here he pleads for setting aside half of earth's surface for wildlife in order to save the entire planet. Reduced ranges diminish biodiversity, and corridors or small plots are of little use. The half-earth equation will create a "safe zone" for all life forms according to Wilson (2016, p. 4). His Chapter 2 is entitled "Humanity Needs a Biosphere," and both humanity and the biosphere have become seriously compromised. While we talk about the climate crisis, Wilson speaks in terms of apocalypse in his Chapter 5, primarily because industrial humans have created a viral synergy. As an example, following Wilson's lead, is deforestation to perpetuate animal farming and growing their feed, which reduces natural habitats, hastens extinctions, minimizes carbon capture, upsets and wastes water, disrupts underground water tables, promotes soil erosion and harmful runoff, and introduces agricultural chemicals and offal pollutions into the air and streams, which deleteriously penetrate larger bodies of water.

Talk of carbon capture technology is for corporations and politicians since forests already sequester carbon. In a list of carbon capture and biodiversity recovery "golden rules," Alice Di Sacco, et al. (2020) place as paramount forest protection. Cal Flyn (2022, ch. 3) makes a good case for abandoned lands to become wild again in order to absorb carbon. She notes how in parts of the U.S., China, South America, and the former Soviet Union reforestation of unused lands is happening. There was an expansion of forest land in the U.S. of 1,400 square miles every year from 1910 to 1979, Flyn notes (2022, p. 75). This is impressive, but these lands need protection and expansion. These numbers do not take into account large tracts of forest already lost. Not only are abandoned lands in America turning wild, but the trend is also happening in Europe, too. This is good news, but there has not been enough in terms of carbon capture. Worse, in tropical countries, according to Flyn, up to one-third of forests are in decline. In its current assessment, the U.N. (FAO 2020) states that forest cover worldwide is decreasing, with a loss of 178 million hectares since 1990. Even assuming a flattening or reversal, consider the amount of natural carbon capture already lost compounded by the continual burning of fossil fuels. Part of Flyn's thesis seems to be that after a country goes through a period of deforestation because of industrial growth to gain economic footing, there are periods where tracts of land are abandoned to rewild where once there were trees, or restored to let nature take its course. In Darwinian terms, species evolve and decline over time; now, extinctions are forced by human degradation and expansion of urban and suburban areas. In Chapter 9, Wilson (2016, p. 71) talks about those who don't preach conservation but cynically say that we simply need to learn how to live on a "damaged planet" and use what's left of the resources as commodities for our comfort.

Human activity and pollutants not only kill plant and animal species but also force others into rapid evolution to select genetic mutations for adaptive survival. Though there is certain death, Flyn's (2022) book demonstrates how ecosystems discover ways to regenerate. That should not grant a pollution license to humans. Other species uniquely adapt to habitats that, once gone, are their death warrant. This dire scenario is why forest sovereignty would wall off ecosystems from foreign invaders that, growing rampant like a virus, will disrupt webs of feeding and networks of water and soils. The fact that humans can inadvertently or deliberately, without understanding the consequences, introduce chaos with non-native plants or unfamiliar animal species into an established ecosystem validates how there's a degree of evolved order in bio-networks that must be protected. The paradox of having interfered with nature means that some monitoring will be necessary. Some newer conservationists want workable forests and grasslands, not fallow ones. Forests are not gardens for human play. Ecological biodiversity is a system of cycles that need to remain intact for a healthy balance.

Rachel Carson and Global Environmentalism

Wilson says that in 1962 Rachel Carson launched a public awareness campaign about the importance of environmental integrity with her book *Silent Spring*. Carson would regret, muses Wilson, to see the endless warfare between environmentalists and "exploiters" (2002, p. 362) who insist on decimating nature with invisible wounds. Even in the late 1950s and early 1960s, Carson laments how, in contrast to the United States, European and Canadian foresters recognized insects and bacteria as part of and crucial to "forest hygiene" (2002, p. 293). Carson's book is filled with examples of how human interference with one plant, insect, or animal species upsets the equilibrium of the whole forest bio-network. Carson's story precedes full-blown climate change but tells the tale of science and technology gone awry. Pesticide companies spending money to unleash deadly chemicals gained the attention of U.S. federal and state agencies controlling agriculture. These actions raise relevant ethical questions about corporate responsibility and the limits of technology to solve climate problems. Seeing the whole of nature, Carson sounds the call to recognize how one dangerous chemical activity destroys biotic communities in a pollution chain. She likens toxic chemicals to nuclear fallout, a war against global nature and a moral question (2002, p. 99), asserting not only human rights but the rights of all living organisms to thrive in their natural habitats.

Carson makes clear that when nature is sickened, human life suffers, too. That statement sounds anthropocentric to gain people's attention. Neither Carson nor most ecologists would necessarily place human life above all else since the planet's energies are in a complex mutual interchange. Carson states diplomatically that "in nature nothing exists alone" (2002, p. 51). Species evolved in response to environments and adapted in biotic communities. Humans are no exception, although we've decided to conquer the planet with population growth that arrests nonhuman relationships and their ecosystems in what she calls our "shotgun approach" regarding flora and fauna (2002, p. 67). Like Darwin, Carson sees the housekeeping economy of nature whereas modern humans try to control nature for their industrial economies in a "habit of killing" (2002, p. 126). Deforestation on scale because of suburban real estate development, logging, mining, industrial agriculture, livestock farming, etc., has created a public health hazard by aiding the increase of pollution and greenhouse gases into the atmosphere while simultaneously decreasing nature's carbon capture abilities. As Carson would say, ecology consists of life and death cycles, naturally. Human interference in forest wildlife, as well as contingent wetlands and grasslands, has disrupted these cycles escalating death. Without forest conservation and preservation all earthly life forms will suffer, unnaturally.

Before Trees, Lichens

Politics are now implicated in species webs, and lichens are examples of organisms on the margins struggling to gain acceptance. The first lichens appeared about 450mya at the beginning of terrestrial ecology demonstrating how neither organism nor ecosystem is fixed but open to dynamic tension. Vincent Zonca (2023, p. 4) says many people see lichens as inferior material rejected from nature. Nonetheless, along with insects, plankton, and mushrooms, he notes that lichens represent 80 percent of living species. Lichens, which do not command the anthropomorphic attention of most people, would present a challenge, Zonca muses, for philosopher Emmanuel Levinas who finds ethics in our seeing likeness in the other. The inability for people to identify with lichens and other such species is at the heart, Zonca suggests, of political aversion to some environmental causes. Lichens, he says, though ubiquitous, are ignored and devalued since they seem to hold no economic worth and are even associated with disease. Zonca's mission is to give purpose and pride to commonplace but unacknowledged lichens in relation to human spaces. This claim is significant, for in understanding the importance of a part (e.g., lichens), people can better apprehend meaning of the whole (e.g., forests).

Lichens are symbiotic. Parts are fungi and parts are algae supporting and inhabiting the whole organism. This pairing of fungi and algae was likely among the first life forms to occupy land (Margulis 1998). Fungi act to root lichen and offer water and nutrients while the algae photosynthesize food as sugars. This fundamental symbiosis of parts, Zonca says (2023, p. 19), explains why hardy lichens universally occupy 8 percent of earth's land. Lichens appear on almost any surface, from volcanic rock to trees, but they need light and humidity, possible even in deserts and glaciers. Lichens act as ground temperature regulators, as microbiomes for small organisms, and as food for many species from insects to reindeer. Zonca says lichens are an important "pioneer vegetation" (2023, p. 21) that build soil and foster other plants in a food web, like mosses, ferns, etc. On stone, lichens slowly break down the skin of rock with enzymes and over time exude elements and minerals to initiate humus for plant growth (Simard 2021, p. 80). Moreover, lichens are a metaphor of life, Zonca says, in their "desire for contact" (2023, p. 35). Lichens are not parasites but in a relationship with their supports, whether trees or rocks, a transitional "second skin" (2023, p. 37) that negotiates internal being with the outside world.

Lichens are multi-dimensional and many colored, not just dull green. Their textures and colors are offset by their stone or wood substrates. As with all living organisms, there's beauty to lichens often lost to most human eyes. Nothing in nature is ugly. Lichens are in a strange dialogue between the living (trees) and the inert (stones) or the dead (fallen trees). Zonca (2023, p. 78) touches on differences of cultures, for example how the Japanese mindset responds respectfully to lichens and mosses in their solitary imperfection compared with other modern attitudes that disdain them. Because lichens can grow in urban places, Zonca suggests we use this indomitable organism to evolve a new and better relationship with nature. Like Jean-Jacques Rousseau, he encourages us not to ignore but rather to study with a new sensibility the smallest aspects of our plant environment to understand how and where we fit into the biosphere, "to relocalize our gaze" (2023, p. 96). This revaluation of perspective includes pre-Darwin eighteenth- and nineteenth-century thinkers who saw God's handicraft in the humble lichen and moss.

As early as the late eighteenth century, but certainly by the mid-nineteenth century, naturalists were beginning to comprehend how through lichens nature was negatively responding to industrial pollution. For instance, it was observed that lichens only sporadically and weakly appeared in the bustling city of Paris, acting as measurements for the hygiene of air. Human development along with climate change might reduce lichens up to 40 percent in the coming years, Zonca fears (2023, p. 111). Urban areas, he thinks, should be reengineered as interiors and not as exteriors of nature. Cities sanitize and clean away from walls vital organisms like moss and lichen that help sustain valuable life forms. He advocates wild spaces in cities and replantation when possible. In fact, he says, the "city is itself nature" (2023, p. 167) and is not antithetical to colonizing microhabitats from lichen to pocket forests helping people reimagine central spaces and their place in nature, an ecology without boundaries. Indeed, to save forests, cities must be built as sustainable natural habitats.

In his special chapter on symbiotic ways of thinking, Zonca notes (2023, p. 174) how fungus "cultivates" alga in a "nutritional strategy" to help both symbiotically survive. We see this symbiosis reflected in many aspects of the natural world, like forests. Thus, lichen raises biological, philosophical, and political questions of individuality; with lichen internally supporting other tiny organisms, it's a metaphor of a forest: a part consists of and contributes to the whole. Lichens, as German and Russian botanists began to realize not without controversy in the nineteenth century, represent plural entities, a form of communalism or mutualism (Zonca 2023, p. 177). Symbiosis, as then coined, meant two organisms cohabiting without parasitism one being of a dual nature, both sharing one life externally and internally. This was confirmed late in the nineteenth century when it was discovered that fungi on plant roots were also symbiotic.

Following a long tradition of thinkers before, up to, and beyond Plato, Zonca (2023) continues his discussion of symbiosis with the idea that sympathetic harmony among species rests on a political foundation. By the 1870s in France, mutualism was viewed as social and biological with voluntary association and shared assistance. Of course, many elements of nature as cooperative were not widely held, justifying capitalistic competition. Zonca cites (2023, p. 188) work of Lynn Margulis who posits that on a cellular level, for instance chloroplasts sharing interacting genetic materials, evolution is fundamentally symbiotic. This idea does not diminish Darwin's notion of the struggle for existence. Links between organisms can become tentative. Zonca notes (2023, p. 194) that symbiosis is not later acquired but is integral to the formation of lichens. Thus, the notion of individuality is held in question since what is considered a single organism is indeterminately open to other potential formations, seen especially in processes of metamorphosis. This is a powerful metaphor that can shape not only ecological but also social thought. Symbiosis is about adding affiliating qualities and capacities to increase an organism's "emergent" capabilities (Zonca 2023, p. 200).

It's not just that genes respond to environment but that symbiotic partnering from the environment can contribute to an organism's genetic constitution. In some sense, many organisms are cultivated in the wild. Although we humans can become a hindrance to symbiotic Gaia, at the same time, says Michel Serres (1995), we can establish a "new social contract" of mutual cohabitation and exchange of benefits with the natural world. As Darwin would say (Costa 2017, p. 94), there's coadaptation of plants coexisting and "portioning resources." Many trees live hundreds of years and lichens, mosses, insects, and microorganisms essential to forest diversity depend on older trees.

Getting to the Root of the Matter

Branching out from remarks in the previous sections, let's look more closely at forests, especially what's underground.

In *Planta Sapiens* (2022, ch. 3), Paco Calvo talks about the dangers of our blindness to plants and how we incorrectly imagine flora and fungi as static background to other forms of life. Contrary to evolution by means of natural selection and descent with modification, the human constructed hierarchy of the great chain of being is punctuated by value coordinates that deemphasize plants. Calvo argues for plant intelligence beyond adaptive responses. He describes how plants store information, communicate with other plants, fungi, birds, insects, and mammals. He explains how plants make predictions about seasons, soil moisture, nutrients, temperature, and light. Through phenotypic plasticity, plant behavior exhibits self-regulation of shape, movement, and internal digestion of water, nutrients, and chemicals. With physiological learning and memory, Calvo attributes psychology to plants in how, he shows, outside of adaptations they actively make choices by using extended cognition. Echoing much of Darwin, Chapter 6 of Calvo's book is entitled "Ecological Cognition" to denote how between a forest being, whether plant or animal, and its environment there's cognitive unity (Tague 2020, ch. 2). This harmony of all living and inert matter in forest woodlands, wetlands, and grasslands is what accounts for the superorganism Gaia.

Land plants play a pivotal role in earth's hydrological cycles not only in transpiring water molecules but also in helping to move moisture and nutrients in soil, says biologist Karl Niklas (2016). Plants are necessary for ecosystems and co-evolved with pollinating insects, birds, mammals, etc. (Crane & Leslie 2014, pp. 143, 149–50). In pollination biology, plants and insects engage in a reciprocal exchange, a coevolution of mutual adaptations by different organisms (Costa 2017, p. 207). With the rise of land plants, animals began to populate earth's surface hundreds of millions of years ago to establish a diversity of niches accommodating worms and snails to insects and mammals (Cartwright 2014, p. 165). Parasitic, mutualistic, and cooperative behaviors evolved in this transition, often referred to as the Cambrian explosion, circa 500mya. As photosynthesis evolved, the atmosphere converted away from other elements to oxygen. In turn, the evolution of plants transformed soil and water movement while providing food for heterotrophic organisms, those who do not produce their own food but eat others. Forest systems with complex cycles developed and transformed land enabling greater evolution of more organisms, phenotypic plasticity, and genetic diversity. Therefore, plant biology is fundamental to all life on earth in fostering complex ecosystems.

Just as genes respond to the environment and pass on fitness-enhancing adaptations, evolving organisms can reciprocally change environments, says Niklas (2016). Plants evolved systems to reward pollinators and seed dispersers. Organisms do not evolve in isolation since flora and fauna occupy a collective habitat. Organisms interacting with each other constitute ecology and evolution, a synergy that once disrupted can devolve into little succession and decline, evidenced in calamitous human deforestation and pollutions. Typically, plant hormones naturally regulate soil microbiomes in dynamic and mutually beneficial ways. An ecosystem of high interconnectedness among species, as in forests, cannot reorganize if it's largely eliminated. Without a home base, there is no exponential growth for most flora and fauna, though soil microbes might survive if not chemically destroyed.

Niklas' mathematical models (2016, ch. 9) show that even small numbers of species can multiply quickly, assuming abundant resources. Presumably that means their habitats have not been dangerously compromised or destroyed. If not killed, animals could possibly migrate to nearby lands if large enough for accommodation, but that's not true for trees. Even under normal conditions there's a carrying capacity to any environs, but that's the point. Left alone, forests are self-regulating and sustainable. Trees live and die in replenishing cycles connected to other organisms. As forests are repeatedly minimized, continuing with models by Niklas (2016, ch. 9), density distribution is unequal, forming a compromised habitat that cannot support growth rates. Important but necessary for restatement, these population declines would not naturally occur because of resource competition but are induced by human causes and in some cases permanent with no chance of rewilding. Niklas outlines survival models (2016, ch. 9) but does not take into account massive deforestation or human pollutions, though there's passing reference to human-caused climate change.

New seedlings need older trees of their species for nurturing root-fungal connections, neural transmissions throughout the forest floor. Suzanne Simard (2021) came to this conclusion. Trees have social relationships where roots spread widely among other plants and function as an intelligent "signaling web" (Calvo 2022, p. 34). With confined areas of cut trees, the potential for natural regrowth is high since seedlings could easily recover. Nursery seedlings not planted in the fungal rich humus but in fully logged areas will improperly root and suffer.

Natural selection favors species who engage in heterosis (de Beer 1965, p. 239), and for trees this means the vigor of cross-fertilization in naturally wooded areas resulting in hybrid superiority (Brewbaker & Sun 1999). As with forester Peter Wohlleben, Simard shows how the life of a forest is not just what we see above ground but consists of the root networks in humus working with bacteria and other microbes, minerals, worms, and especially fungal varieties. Fungi are essential to ecosystems and the carbon cycle as they break down plant biomass (Hibbett 2014).

Forest recovery needs more than just seedlings. A foundation of established trees and other living entities dependent upon each other is required. Mycorrhizal fungi underground supply water and nutrients to the tree through its roots in mutual exchange for sugars from the tree's photosynthesis (Simard 2021, p. 60). This cooperative relationship benefits the tree and saves it the cost and trouble of producing more roots than necessary since the fungal networks can reach nutrients far and wide. Simard makes clear that in her years of forestry while working in British Columbia, Canada, circa the 1980s, logging companies under the direction of the government were unaware of the damage they were inflicting with clear cutting and feeble efforts to restore old-growth forests. Across timberlands globally, larger, older trees "fix" carbon reservoirs and are vital to stabilizing optimum atmospheric conditions (Stephenson, et al. 2014). Old-growth forests healthily connect ecosystem organisms (Gilhen-Baker, et al. 2022). Simard's research (2021, p. 96–97) suggests that mycorrhizal fungi are symbiotically shared between tree species. More so, the healthy growth of trees depends on the right type and variety of soil fungi. She also discovered how alder shrubs, thought to be a forester's bane, collaborate with pines by adding nitrogen to and rehydrating soil later in the summer. This activity proves how native species evolved to benefit each other in a Darwinian synchrony of variation, selection (cooperative as well as competitive), and retention. The pines nourished by the alders through soil fungi thrived, and the added strength helped fend off parasites. In time, Simard's research would reveal how trees relate to each other through fungal root systems. Her work supports a whole forest approach, similar to the Menominee Tribe of Wisconsin foresters, U.S., who methodically log dying trees while preserving elder trees that are thriving (Buckley 2023).

Weeding ruins and erodes top soil, robs the earth of nutrients and the microorganisms that create humus, and spoils mycorrhizal fungi diversity vital to tree root systems. Sunlight through photosynthesis equates to the energy of carbon, hydrogen, and oxygen and works its way as sugar or as broken molecules to help the roots spread and then leach out to other trees. Cutting down woodlands harbingers bad, unending effects on the environment. Not to oversimplify her work, Simard (2021, p. 134) talks about how removing or limiting the growth of birch and aspen negatively impacts conifer health through root infection and disease. A young birch photosynthesizes in abundance and pumps excess sugars into the soil. In summer, birch trees provide more sustenance to fir trees than they receive in return through a language of carbon transfer via root fungi. During the fall and spring seasons, Simard goes on, firs aid birch more, with both species creating a neighborly reciprocity. Further experiments by Simard (2021, p. 254) would show interspecies nutrient transfers and the exchange of stress signals. Tree reciprocation is real and operates underground. This important conclusion shows how the elimination of weed trees was detrimental to a healthy forest. Additionally, cutting away thousands of trees unnecessarily and disturbing the soils releases captured carbon and prevents future carbon capture.

In other experiments, Simard (2021) discovered that established trees overshadowing their seedlings and depriving them of sunshine nevertheless communicate and cooperate through fungal connections in linked roots. Still, Douglas-fir seedlings simultaneously reach out to birch, establishing a two-pronged mycorrhizal fungal web. There are many varieties of mycorrhizal fungi serving different purposes and needs in older forests and why preserving those soils is vital for elder as well as fledgling trees. Despite all of this mutual aid one can't discount competition, as even Simard would admit. Trees compete with other organisms, too, but they have evolved adaptations for survival, unless cut away, burned, or hampered by climate change. The fungus benefits from working with different species, as well, spreading risk in a competitive altruism to gain carbon energy. Musings like these lead Simard (2021, p. 187) to speculate that above Darwinian individual selection there is probably group level selection where communities of different species collaborate and compete adaptively to increase fitness resilience for survival and reproduction. More mixed forests with native species, not monocultures, are advantageous for species biodiversity and climate health by providing helpful bacteria and microbes. Selection favors polyculture, not monoculture (Costa 2017, p. 99–100). While climate change is weakening trees and helping parasites survive longer, Wohlleben (2019, p. 83) says a large part of the problem is monoculture tree plantations, enticing bark beetles with tree food. Under normal circumstances the beetle preys only on weak trees.

Simard (2021) shows that an elder "mother tree" can be linked to dozens of same-species seedlings through mycelium or fungal fibers in a far-reaching underground tapestry. These webbed links are mimicked by other species like shrubs and berry bushes. This parent-offspring hydration and nourishment in molecular transmitters, which can only happen in a forest and not in a plant nursery, sounds like kin-selection theory, recognized by Simard (2021, p. 269). Forests need to evolve as united, fully functioning superior organisms at their own pace and should not be planted by humans, so says forester Wohlleben (2016). He notes how managers in Yellowstone National Park, U.S., echoing Leopold (1949), killed most wolves in the 1920s. This led to an explosion of the elk population, which meant aspen, other saplings, and forest vegetation were decimated through browsing. When the wolves were allowed to return, so did the trees, balancing the ecological system of nature's ways. Caroline Fraser (2009) also recounts ecosystem decline in Yellowstone after wolves were removed. With browsed aspen there was no building material for beavers, so those populations declined, which meant there were no large pools for water plants, birds, and other wildlife, like fish. Wolves also control and restrict beaver feeding ranges resulting in more concentrated wetland areas (Bump, et al. 2022; Gable, et al. 2023). Some researchers (Hobbs, et al. 2024) argue against riparian ecosystem recovery through the reintroduction of apex predators since Yellowstone remains, because of human intervention, in an "alternative stable state." This estimation portends the need for solitude and evolutionary time so flora and fauna can regain their optimal balance.

A forest landscape needs all native flora, fauna, and fungi to flourish. Trees near streams cool the water to maintain optimum temperatures for trout. In other words, trees are not merely lumber for people, which grossly narrows a forest's function and utility to human desires. Trees should be allowed to die in place, engaging the corpse with relatives in carbon and nutrient exchange through root interconnections (Simard, et al. 1997; E. Fraser, et al. 2006). Much of this activity is vital in terms of capturing and storing carbon dioxide from the atmosphere, a crucial concern because of human-induced climate change. There is interdependence of roots among trees of the same species. Furthermore, interspecies communication through root fungi to facilitate forest survival attests to Gaia's self-regulating systems without human technology (Y. Song, et al. 2015). Fungi provide phosphorous and nitrogen for plants; via photosynthesis plants generate sugars for fungi (Calvo 2022, p. 39). Relatedly, conservationist Bernhard Grzimek (Attenborough 2020) studied how wildebeest and grasses beneficially co-evolved, as one instance of that natural phenomenon. In a similar way, the migratory feeding movements of Yellowstone bison engineer green waves of springtime vegetal growth (Geremia, et al. 2019).

Let's broaden this holistic study. The science confirms that trees communicate with each other through vast fungal root system networks about the weather as well as insects. Above ground, trees use sweet scents and sugars to attract pollinators. These natural systems have been affected by human disturbance. The viability of each tree is dependent upon other trees nearby, so culling trees to open a canopy space deprives them of sharing resources or managing genetic diversity at their roots. In the life cycles of trees, at times they need to propagate by growing seeds, which take up vital leaf space on branches resulting in less photosynthesis and hence less sugar. This means, says Wohlleben (2016, ch. 5), that trees become vulnerable to pests and harmful fungal disease. Some trees emit a natural pesticide, phytoncides, as a defense against predator pests. Using statistics, Wohlleben recounts how each tree over its lifetime successfully produces only about one offspring in spite of many seeds it might have dispersed over the years. Seeds that don't germinate are not wasted since they are eaten by animals or decay to become humus components.

Wohlleben (2016, ch. 6) points out that trees grow slowly and can live for a long time. Slow growth is important for proper ripening of strong, dense wood and defense mechanisms against insects. Even mother trees know this by shading offspring from too much sunlight. The young trees must wait their turn as older trees die and open up space for important sunlight, all the while fending off hungry animals and meddling vines like honeysuckle that can choke growth. Small, narrow branches, says Wohlleben (2016, ch. 7), are replaced by larger ones as the tree shaft stiffens and grows upward, developing its crown and stability. Water supply is a critical need for trees to help them produce and circulate food for growth, generated by photosynthesis. Trees learn, he insists, to ration and store water and not squander what's available, which could cause a sudden shock that splits and dries bark. Trees compete for water, and in cases of beneficial fungi, tree roots capitalize on the opportunity to extend their water-thirsty roots through the mycelium or fungal vegetation (Wohlleben 2016, ch. 9).

This is cooperation. The tree does not kill the fungi on which it depends and with whom it shares precious food like carbohydrates and sugars. These helpful fungi can extend to other tree roots for even more sharing of water and nutrients. Wohlleben (2016) says that this fungal network is also a means of communicating attacks by insects and warding off destructive bacteria. While trees, from their tops to their roots, can be competitive, their various fungal parts help them all survive, perhaps because they need multiple hosts. Water rises up through trees by capillary action and transpiration through leaves. Trees exhale their moisture, which creates a force of suction that can pull bonded water molecules from the roots up through the very narrow capillaries. Even still, Wohlleben mentions that a third, theoretical mechanism for upward water motion is osmosis in roots and leaves with fluid moving between cells and filling empty pockets to create a balance. In some places algae or moss grow on old-growth trees, and the capture of water and nitrogen high above trickles down in rains to fertilize the forest floor.

Wohlleben (2016) says there's really no ideal place or conditions for a tree to grow, contending with all kinds of opposition forces. In terms of competition, he describes (2016, ch. 12) how a young beech can penetrate the root system of an oak and sap all resources. As the beech grows taller than the oak, its leaves overtake the oak's crown for sunlight. Otherwise, oaks are quite strong and can live up to 500 years. Rather, it's more about the tree adapting to a place than a perfect ecological niche, though that concept applies. Some trees are better adapters than others, like spruce. Then some others, like yew, develop superior root systems capable of producing multiple trunks to compensate for slow growth and the prospect of damage above ground. Adapting to forest conditions in this way, a yew tree can grow to an advanced age of 1,000 years having accumulated most of its growth under low-light conditions in the shadows of other trees it will outlive. Unlike most other trees, alders, for instance, can grow and survive in wet, swampy soil. Roots are the primal and vital life force of a tree, which store memories, nutrients, and promote growth. One spruce root system in Dalarna, Sweden consists of expansive ground shrubbery dated to almost 10,000 years of age.

Root systems have electrical signals that respond to stimuli, affecting behavior and movement in the soil toward nutrients and away from harmful substances, says Wohlleben (2016, ch. 14). Echoing other scientists like Niklas (2016), he claims that large-bodied animals and even birds are less crucial to the forest ecosystem than the soil and what's in it, for example fungi. Birds, however, can transport microscopic organisms to different locations, as even Darwin speculated. Furthermore, tree roots stabilize the soil and prevent it from eroding. Without trees, a compromised forest can lose almost 3,000 tons of soil per square mile each year through erosion. Soil is enriched not only by beetles, weevils, mites, spring tails, and pseudo centipedes but also thousands of other microscopic beings. These tiny creatures act as eco-engineers, turning forest material, from dead tree limbs or leaves and animal corpses, into the humus from which trees can root. Many of these small species are in decline, since forests are cleared and the floor becomes empty, losing crucial "nutrient cycles" completed by fungi and bacteria (Wohlleben 2016, p. 91).

Many hundreds of years are required to regenerate healthy, diverse forests from soils to tree canopies, and this renewal requires vast expanses free and protected from human intrusion. Old-growth forests were once abundant in Europe, but no longer. An average tree can store up to about 20 tons of carbon in her root system and body. Wohlleben (2016, p. 93) refers to a forest as a "vacuum" that is always filtering and storing carbon dioxide. Additionally, he notes large trees can filter up to 200 pounds of atmospheric dust a year (2016, ch. 26). Some CO_2 returns to the atmosphere when a tree dies, but most CO_2 is stored for eternity in a forest ecosystem. A tree that dies naturally in the forest is broken down over time by natural forces and becomes particles contributing to the humus, storing compressed CO_2. Scientists (Gauci, et al. 2024) also disclose that tree bark absorbs the greenhouse gas methane, a harmful contributor to global warming.

During the summer, 29 tons of oxygen are released by trees in a square mile; someone needs about 2 pounds of oxygen per day, so a square mile of forest supplies oxygen to about 10,000 people. At night, since they are not photosynthesizing and rather converting those sugars to energy, they release CO_2, which only means a decrease in oxygen level. During winter without leaves, microorganisms are breaking up forest floor fodder producing CO_2, but because of global air circulation and algae producing oxygen in water, there's a balance. Soil carbon sinks work best with steady plant growth. At the same time, this means grasslands can become the new areas of carbon storage, where trees should not necessarily be planted (Garthwaite 2021; also see Silveira, et al. 2012). While planting new trees has value, older trees in ancient forests are the true climate champions and should be left alone to do their work. Trees benefit their habitats by leafing the ground to hold moisture, extending branches that cut down harsh winds, modulating forest floor temperatures, and creating humus cover that promotes micro-culture growth.

Spruce and conifers, which prefer cold regions with melting snow, never really adapted to retaining water like the deciduous trees (beech). Tree canopies, in square footage larger than the forest floor, capture and evaporate moisture. Trees transpire back into the atmosphere water stored underground. All of this is important since the atmospheric moisture released by coastal forests creates rain clouds for tree areas inland. This explains why rainforests, especially but not exclusively, are so important in terms of pumping life-sustaining moisture into the air. Without coastal forests, the water pump stops. Conifers emit terpenes to ward off pests, but this chemical helps thicken clouds to reflect sunlight and increase rainfall. These items are further proof of how forests can fight climate change, says Wohlleben (2016, ch. 18), who goes on about how trees aid in setting up the forest floor as a saturated reservoir, and water collected there is important for maintaining a biodiversity feedback loop. In this way, trees facilitate small streams and pools, important habitats for many creatures vital for forest health.

While he speaks of trees communicating and sharing, for other organisms Wohlleben (2016) addresses the struggle for survival where each life form takes something from another. At the same time, while species engross all they can for survival, there are checks in place to prevent forest collapse from exploitation. For instance, various animals and birds eat acorns and nuts but bury or drop some, which can germinate or supply food for others. Trees are food: sugar, carbohydrates, cellulose, lignin, and water for species from woodpeckers and aphids to bacteria and fungi. Simultaneously, there are organisms that eat aphids or tree honeydew they discard. In the latter case, Wohlleben continues, ants will actually protect the aphids so as to have a supply of food. Fungi will make a home of honeydew on the trees, and bees feast on it, too. A large population of caterpillars, however, can denude a tree, and there are other species that burrow into the tree bark and dine on cambium in secluded protection as free riders, sometimes containing dangerous fungi. Birds defend tree leaves by eating insects (Van Bael, et al. 2003), and trees guard themselves with secretions, some of which attract wasps who will eat the invaders. Deer and other ungulates rub bark from older trees or eat the summer bark and browse juvenile trees. Fungus and mushrooms with creeping mycelium overtake roots of trees and grow into the bark penetrating the cambium.

Nonetheless, this orchestral dance of struggle for survival and reproduction is actually what keeps the forest alive and healthy, as it has been for many millennia before disturbance by modern human machines. Forest trees, it should be obvious by now, have evolved adaptations to work reciprocally with countless other organisms, often in tandem to drive away as many invaders as possible. On the other hand, forests managed for lumber grow trees quickly, which means their nursey roots are sparse, their wood is less dense, and they are more disease prone. One study of forest plots in Germany (Boch, et al. 2013) indicates that overall plant richness can benefit from human management. However, the forest studied is used for timber production and has been for ages, so it is not pristine but degenerated, hence requiring assistance. Numerous decades must pass before a compromised tree dies, and when it falls light opens for others. As a tree decays, its body becomes a feast of nutrients for insects and other chewing organisms breaking down all the parts into humus. Nothing is wasted but instead is recycled. In the forest, feeding is a sustainable behavior that keeps population balances in check. Scientists would agree that all of this diversity acts to stabilize forest ecosystems. This means many species combined do not permit any other to stage a takeover, excepting humans. Most species don't harm trees and simply take advantage of a safe haven of light and moisture. Through a process of "cadaver rejuvenation" (Wohlleben 2016, p. 135) young trees benefit from and grow in a deceased parent until they root in soil. In all, Wohlleben corroborates the Gaia hypothesis of Lovelock (1979), particularly regarding forests as self-regulating systems.

Many species, in contrast to trees, reproduce in short intervals, which is good for gene shuffling to aid in adaptations as the environment changes. Trees, however, live on average for hundreds of years. Trees can tolerate abrupt climate changes, evident in some species rooting from the Mediterranean to far northern climates. Different genes activate in line with the environment early in a tree's life, but adaptation occurs in older trees, too, whether in response to pests, heat, cold, or water conditions. Yet a new generation of trees does not happen quickly, so transfer of adaptations takes time. In a forest left alone, microclimate genetic diversity among species is large, increasing chances for the passing of beneficial mutations. This is basic Darwinian evolution: variation, competition, and inheritance. The problem is that machines damage land in many ways by disrupting wildlife, altering water tables as food systems for forest organisms, adding chemical pollutants, and compacting soil so trees cannot root. Forests don't stand still but move, including the migration of trees, albeit slowly. People, too, can introduce new species, some of which can be harmful. Either way, this change of environment can be a gamble. As Darwin knew from his experiments with plants, trees are not inert but respond to light, water, air, other trees, organisms, and temperature variations; they grow and move around though activity underground is faster. At the time Wohlleben writes (2016), there was consensus among German politicians that 5 percent of existing forests should be left to their own devices, what is called process conservation to ensure old-growth forests for the future. This tiny percentage is probably not enough, given how humans continue to mine and burn fossil fuels. Ironically, some environmentalists and local communities resist wind and solar farms as a form of encroaching development or land loss (Tankersley, et al. 2023). Wohlleben argues for a policy of non-interference, noting that it can take up to 500 years for a forest to reclaim a natural state. Common sense and the evidence in this chapter support a policy of expanding forests, including their adjacent wetlands and grasslands, and preserving them and protecting Indigenous rights with a hands-off approach unless intervention is biologically and ethically necessary.

Neil Evernden (1993, p. 4) proves that from the 1980s, if not earlier, there was concern about the environment from political and cultural perspectives as a pressing social movement for environmental advocacy. Evernden (1993, p. 8) says forests can be protected by "benign neglect" and argues against nature as a resource center for capitalistic expediency. Working with governments, manufacturers and developers gain from environmental impact studies since the calculus does not benefit nature but capitalists, according to Evernden (1993, p. 11). He validates intrinsic value in nature, important for the argument unfolding here. Readers see by now how there are differences between an untouched forest and one intruded upon by humans. People on every financial level and of each cultural background should not detach from and objectify nature to deprive flora and fauna of life. As Kropotkin, Leopold, E. O. Wilson, Zonca, Wohlleben, and Simard reveal, mutualism, symbiosis, and cooperation as interspecies efforts help forests flourish as a basic factor of Gaia. Science principally asks *how*; the more crucial question is *why*. In line with Lovelock and Margulis, we know how nature works; why do we mostly continue to separate ourselves physically, emotionally, and ethically from the natural world and mechanistically view forests only as a commodity for human exchange.

In the next few chapters, we'll see that at its source political philosophy has ignored lessons from woodlands but obliquely through my reading supports a theory of forest sovereignty. Since this material, though selective, is dense, I've broken the analysis into several chapters. Then, we can circle back to the place of humanity in nature before concluding with an ecological credo.

Chapter 2

Early Political Theory and Implications for Forests

In this chapter, I'll weave the discussion of key theorists into an argument for the sovereign forest state of Gaia. The focus is on, but not exclusively, key writings by Thomas Hobbes, John Locke, Jean-Jacques Rousseau, Adam Smith, John Stuart Mill, and Karl Marx. Since this is not a political text, I have made no effort to cover all philosophers but only those who seem salient to the discussion. Regardless, not one of the thinkers outlined hereinafter has expressed true concern for nature but states or suggests natural resources exist for human use alone. In the Preface I offered a proposal, and it's important to note how the critique of Western, Enlightenment political philosophy calls for action. The voices and ideas across a diversity of underprivileged and underrepresented people committed to wildlife and climate health need to be considered in planning and development that impact forests, wetlands, and grasslands.

Jonathan Wolff (2006) asks about fair distribution of property given the rights people should have. While people certainly enter into the picture, a more precise question hovers around property rights, distribution of property, and political power for animals in forests. From a human perspective, there could be power in nature; from another perspective, it's all about energy cycles. There's homeostasis of liberty and justice in nature, all things considered, but a state of Gaia is necessary to protect forests from humans. People are political, evidenced in how nations purportedly need institutions more than wildlife. Without political influence human life might look like the symbiotic relationships and mutual aid we find among forest creatures. Wolff (2006) asks if humans can only live in a civil society. That question props humans and "humanity" above all other life forms and ignores our hominin predecessors who lived a life in groups similar to those of hunter-gatherers. A forest society is a combination of property holders (via Locke), consumers (via Smith), and producers (via Marx). As Peter Burdon (2015) emphasizes, there ought to be an eco-centric notion of earth jurisprudence that faults private property as anthropocentric and aims to center nature in the legal system. If Gaia can be seen as an ideal society, it's often broken when capitalists control it as private property, according to Plato's *Republic* (Hadas 1950, p. 139). Burdon is critical of human laws that marginalize eco-centric ethics over property interests. From another perspective, sovereignty need not reflect human law but the laws of nature. Private property owners could be incentivized to keep their lands green.

Recalling the aim of this book from the Preface and what was learned in Chapter 1, you might ask about any link between political philosophy and forest sovereignty. On the one hand, political philosophy has been used to rationalize harms to nature and why there should be a protected, sovereign forest state. On the other hand, much political philosophy can be used as a lens to justify forest sovereignty. As outlined in Chapter 1, and echoing American Federalist paper number 49 (Bourne 1901), no power is surrendered in a union but distributed. The Federalist writers say that sovereignty is independence "bound by its own voluntary act" (number 39), essentially the argument here about wildlife plants and animals. There's unity of multiple interests, with no "incompatible alliances" (number 7). Overall, John Jay, Alexander Hamilton, and James Madison try to prove in their Federalist papers that there's no single sovereign but an integrated collection of autonomous individuals and units comprising one union. The attempt in employing political philosophy is to illuminate the inherent value of forest, wetland, and grassland unity justifying biological preservation and perpetuity as an independent wildlife state. Since much philosophy, especially in the Western tradition from which this discussion draws, has been prejudiced by religion, we first need a section to make some distinctions.

Philosophy and Religion

As of this writing, there are still many colleges that integrate into one department philosophy (knowledge and reason) and religion (faith and worship). Not unrelated, in almost any culture one finds images of deities in art, song, and narrative from ancient Egypt and Indigenous people up to the modern era. In some ancient cultures, there were of course multiple Gods. According to Bruce Clarke (2020, p. 60) invoking Bruno Latour, Western thought conflates God and nature, confusing two ideas. Even so, it's unfortunate to some degree that Clarke and the others he discusses take the science of Gaia and play it into something allegorical, metaphorical, semiotic, and posthuman. Given human language, myth, and ideas, this approach is understandable but continues to separate us from the reality of a nonhuman living world. Gaia as an evolving order is not equal to God or a soul. Semantic hair-splitting about what Gaia is/not, particularly when it leans into religion, doesn't really support the proposal outlined here.

Generally speaking, Eastern religious beliefs might differ from others in that everything in nature including the human animal is connected, as evidenced in parts of this book. There are diverse Eastern traditions with sophisticated views on the sanctity of nature. Since modern, industrial nations, and hence regions they've colonized, have mostly relied on Western philosophy and religion, that's what's discussed. Nakamura Hajime (1992) in comparing Eastern and Western ideas of nature, talks about the Buddhist satisfaction with what is, a quieting of the will. That position has not typically been present among modern, Western nations, nor the idea that people and nature are one and not separate (Lee & Son 2016).

In 1387, Lithuania was the last pagan culture in Europe to be Christianized, and we see in Chaucer's Knight from *The Canterbury Tales* (circa 1400) how that warrior was part of the violent campaign. Pagan cultures, like Lithuania's, evince rich, mythic explanations of nature, earth, and its inhabitants by evoking deities. Overall, this basic idea of the interconnectedness of life and natural elements is reflected in what you've read so far and is important to remember while reading the following sections in the next few chapters. One of the first modern Western philosophers, Arthur Schopenhauer (1788–1860), embraced Eastern thought. On the other hand, it's difficult to read Manfred Kuehn's (2001) biography of Immanuel Kant (1724–1804), who immensely influences ethics to this day, without encountering references to the Christian God. Ineluctably, one finds God in philosophy from the ethics of Emmanuel Levinas (1906–1995) to the existentialism of Paul Tillich (1886–1965), and everything in between, including modern Eastern writings on ethics like the Buddhist scholar D. T. Suzuki (1870–1966). Friedrich Nietzsche (1844–1900) famously cried that God is dead, but what he really meant is that faith in God has been killed and that a godless world is like apocalypse (Kaufmann 1974, p. 97). In *The Varieties of Religious Experience* (1961), psychologist William James (1842–1910) wrestles with the absurdities of religion that somehow functionally operate in modern life. None of this religious belief explains the workings of a forest sovereignty theory as presented here; the two are separate.

Natural philosophy, the precursor of what's called science, often invoked God though it sought materialist explanations of the physics and life cycles of nature. We can see this temperament in one of natural philosophy's early proponents, Francis Bacon (1561–1626), who developed what's called the scientific method. The blending of this early science and religion continued through the seventeenth and eighteenth centuries. The Victorians of the nineteenth century and beyond, though living in an age of speculation and doubt after the publication of Darwin's *Origin of Species* in 1859 and especially *The Descent of Man* in 1871, were characteristically of a Christian mindset (Houghton 1957, p. 21). Darwin himself, though a naturalist, studied at Cambridge University in order to enter the clergy after abandoning medical studies in Edinburgh. At Christ College, Darwin, like all students of the time, studied William Paley's natural theology drawing from a tradition going back to the sixteenth century. Simply put, Paley's argument was that everything on earth is specially created by the Christian God, and for the most part, Darwin found this approach "irresistible" (Browne 1995, p. 129). Later, after he developed along with Alfred Russel Wallace (1823–1913) the theory of natural selection, much opposition to Darwin's work came, not surprisingly, from religious leaders. Darwin himself carefully shrouded his agnosticism to avoid shaming his devout wife. Religious thought attempts, in one occurrence as a prominent example, to meld quasi-Darwinian thought with intelligent design in Henri Bergson's (1911) notion of *élan vital* or creative evolution; regrettably, in some ways this approach is a step back into natural philosophy.

More recently and perhaps a bit closer to thinking in this argument, Thomas Berry (2014) asks for Christian responsibility toward the environment. The caution is that Berry advocates for the putative religious implications of Gaia. It's important to note that Berry (1914–2009) was a scholar and teacher of Eastern philosophy and religion as a Catholic priest. In spite of his Eastern bent, he relies quite a bit on Saint Aquinas, Saint Augustine, and scriptures. Berry is on point to assert how religions have essentially ignored ecology; he wants to move beyond religion but repeatedly speaks in religious terminology. Berry is not exactly irreligious, but he often uses words like divine or divinity, sacred, numinous, spiritual, soul, spirit, religion, or creator (implying a Christian God). Berry's contribution envisions an Ecozoic Era where humans would once again organically align themselves to earth. While this notion took some hold from the 1970s to the 1990s when Berry was flourishing, it's now hard to imagine in a more secularized world dominated by computer devices, video games, cell phones, electric vehicles, artificial intelligence, etc.

Berry (2014) speaks in language of cosmological mythology and then about the human place in evolution. Myths cannot replace science in earthly restoration, as he suggests. Those two are not compatible. Nor is creative evolution a science, as he hints. Often, he returns to the question of our role in nature, which reverberates with anthropocentrism. Indeed, he believes we were chosen by divine power. He might be right to suggest that humans are, in some primitive way, a self-reflection of earth. Berry is more toward communion and not differentiation, but that idea fundamentally goes against evolutionary processes. At points leaning toward intelligent design (pp. 117–19, 124), he seems to fight against random variation in nature, the building block of evolutionary descent with modification.

Berry's (2014) thinking floats around what he calls the "divine" in nature. He wants to avoid saying "God" by invoking science, but in the next breath he will use a word like numinous. While biologists including James Lovelock and E. O. Wilson might talk of earth as a superorganism, Berry paints earth's life as super eminent. What does he mean? He thinks humans have enabled earth and assumes humans could have been earth's glory, but that's stretching into the false idea of human exceptionalism. In contrast, the Puritan and Bible scholar John Milton, in *Paradise Lost*, makes clear that Adam and Eve are expelled from Eden for having polluted the earthly paradise. While Berry wants to veer away from anthropocentrism, he often collides with human-centered thinking. Worse, he labels Darwin as "numinous" and the universe as "sacred." His belief is that we need to recover relationships with animals and not be averse to wilderness, but then goes on to propose how human interference in nature calls for yet more intervention, the opposite of what's proposed here. At the same time, Berry is aware of how the doubtful divine ordination of Manifest Destiny destroyed Western America and Indigenous ways of life and how a writer like Aldo Leopold focusing on a land ethic is still quite relevant. Nonetheless, Berry often slides into spiritual or religious explanations without any science to discuss a "sacred world" (p. 53). Some believe Berry operates in Gaia theory, but that's difficult to see.

The bottom line for Berry (2014) is that the natural world is a source of spiritual consciousness. As good as that sounds, it does not precisely translate into action. Specifically, he talks about religious experience via the universe, but one wonders how that will lead shrinking congregations toward ecological conservation as they meanwhile embrace technological gadgetry. Adversely, how will religious faith and notions like Isaiah 11:6–9 with God as the great engineer help us move forward and embrace science? Berry talks about an "integral ecologist" (p. 83) or a blending of science and traditions. Some traditions are unethical (patriarchy) or unhealthy (animal agriculture) and must be divorced from forward-looking scientific thought. Some of the malevolent ideology in religions stems from ideas like "thy will be done"; rather, in order to fight human-manufactured climate change, we must exercise our collective and scientifically informed will to engineer an ecological ethics without religion.

Let's see how this section plays out in the discussion of political philosophy in this chapter and the next where many of the philosophers, excluding Marx, either grounded their ideas in a Christian God or at least hold some religious beliefs. Thomas Hobbes, for instance, seems to rely less on God and more on a first mover. This section is important and has been strategically placed to help readers understand how religious, mythological, and spiritual beliefs are ingrained into many cultures; human tendencies to explain the natural world and our place in it without biology or physics could inch our way forward, as attempted by Thomas Berry. At the same time, unscientific opinion can also become a drag on movement toward forest preservation and a cleaner environment. There are many Indigenous people with diverse beliefs, but it's fair to say they have a closer connection to the natural world through animal and ancestral spirits, shamans, and a range of deities. For more on Indigenous cultural concepts of nature, the reader is directed to the section Indigenous and Global South voices in the Conclusion.

In reading the following sections, since this work is cumulative, keep in mind the general observations and science from the Introduction and Chapter 1. That is, try to read the great philosophical minds on human rights and privileges from the standpoint of wildlife.

Thomas Hobbes (1588–1679), *Leviathan* (1651)

R. S. Peters (1967) says that Hobbes believes in the absolute sovereignty of a leader without relying on the standard argument of divine right, though the Judeo-Christian God looms in the background dependent upon the sovereign's beliefs. Indeed, Hobbes tries to suggest how any social contract and representation calls for a strong ruler. In sum, contrary to the perceived chaos of nature, for Hobbes the office of the sovereign is authorized by a covenant of those governed, so somewhat liberal; yet the sovereign has unlimited powers with the aim of "social equality" (Smith 2021, pp. 156–60). In truth and differing from Hobbes' philosophy, left unscathed the state of nature is in balance without a supreme leader and needs no remedy; human prescriptions have caused the apparent disorder there. Hobbes addresses the question of where an individual fits into a nation, compared to the church-dominated states of the Middle Ages.

Hobbes imagines a new world order of socially mobile individuals in competition, which raises the specter of anarchy as he claims exists in nature. Hobbes sees control stemming from reasonableness of the state, laws, and conscience, inner regulation and not what was epitomized in the Roman Catholic Church. Outside of the human perspective, wildlife is not about power and control but connectivity and self-management. Social contract theory is used by those contesting the authority of a monarch; meanwhile without any ruler, plants and animals in the wild have evolved a form of social contract in mutual aid and symbiosis. Generally, competition in nature is to evolve heritable adaptations, not to exert dominance over others.

For Hobbes, it's not government by consent but for security. People are driven by self-preservation and self-interest, and without laws or authority there's chaos. A state is justified if it protects the lives of citizens even while denying them freedoms, as Hobbes doubted democracy is "stable" (Christman 2002, p. 34). There's no social compact for Hobbes since that's a third power above both citizens and sovereign. The sovereign state for Hobbes rules in a contract for the overall interests and not the individual will of citizens, whereas for Locke it's will and not rights (Christman 2002, p. 36). Hobbes causally links law and authority. Natural rights are not just entitlements but those someone cannot psychologically avoid, like self-preservation or fear of death. This psychological realism is apparent in wildlife, too. Hobbes' notion of natural rights differs from Locke's, who sees natural rights as protections from others. As for the commonwealth, Hobbes says mere covenants are empty words if they don't have the force of law in a powerful sovereign, though people have representatives who make ruling decisions with them. Security, arising from fear of death might not be the sole rationale for a commonwealth. As Locke notes, absolute rule is dangerous. For this discussion, there are questions about the validity and equity of human laws that harm nature with detrimental consequences to people worldwide.

Like John Locke after him, Hobbes places much emphasis on the reasoning capacity away from carnal pleasures of, as he says, "man," in contrast to animals. Of course, much scientific evidence has demonstrated that animals have the ability to think and act (Jamieson 2018). At the same time, Hobbes finds in "mankind" an indefatigable desire for power. Specifically, Hobbes says that it is human nature to compete with others and invade their lands for gain, safety, or glory. We do not see these tendencies, typically, in animals, even putting aside any accuracy in Hobbes' claim. In a constant state of conflict, there can be no industry, says Hobbes, concluding with his famous expression that life is then "solitary, poor, nasty, brutish, and short" (ch. 13). That condition would be what humans inflict upon themselves; but we don't see that in nature where many creatures exist in ecological niches that interlock and support one another perpetuating the industry of nature. What we see as injustice in nature, such as a big cat preying upon an antelope for food, is an evolved feeding behavior, not a malicious choice.

Hobbes had such fear of civil war, as he experienced in England, that there was a call on his part for a strong, authoritarian leader. There must be a common power to enforce laws and establish justice. He sees justice only in human society (ch. 13). He agrees that in nature we find moral virtues, but they are not in themselves laws and can only be so in a commonwealth (ch. 26). Besides, like other thinkers of his time, "in nature" meant Indigenous people and excluded animals. Later in the seventeenth century, Locke defines commonwealth as an "independent community" (*Second Treatise*, ch. 10), much as we see in forests. Hobbes does seem to admit that animals can live together, as he surely observed. The difference is that humans live together by a covenant for a common benefit. We know this is not true since there is so much greed, destruction, conflict, and warfare among people, as Hobbes would be the first to admit. Hobbes has a disproportionately negative view of the state of nature, though our ancestral hominin relatives and hunter-gatherers, to say nothing of primates and mammals, managed to live peaceably enough to survive for vast stretches of time. Hobbes believes in equality among humans in nature, but put alongside the scarcity of goods, desire erupting from individual and group power would lead to conflict. For Hobbes, this state of affairs means that people are always ready, willing, and able to fight. In the natural world, physical battle is often avoided since there's danger of pain or injury.

Wolff (2006) claims that the human moral sense can be a corrective to any warring inclination. He goes on, like Hobbes, to say this moral sensibility is not evident in other creatures. In contrast, see Victor Kumar and Richmond Campbell (2022) who, though they privilege "humanity," find the elements of morality in many animals. Biologist Richard Alexander (1987) more convincingly discusses the biology of moral systems in evolutionary terms. Studies show how animals exhibit fairness, empathy, caring, non-reciprocal altruism, etc. (Monsó, et al. 2018). There is no outright combat in nature, except from an anthropocentric view. Instead, each creature has evolved adaptations to fit into an ecological niche and survive differently with due consideration for others in a shared environment. Hobbes says free conflict is a natural right, but he's likely thinking of the European wars of the seventeenth century among developed nations. Hobbes sees people following the laws of nature as self-preservation. That's what animals do, often at their peril under the rule of human hands.

Contradictorily, Hobbes talks about a rational peace in nature's state of war. In explaining this, Wolff (2006) discusses individual and collective rationality. Individually, people will be self-interested to the point of conflict with others. Collectively, people will act with self-interest assuming others act likewise. Coming to grips with that ambiguity is part of our animal nature. In this way, Hobbes expects others to obey laws and not exploit the system, but some people do not act morally putting them in conflict with conformists. For Hobbes, the solution is a powerful and all-punishing ruler. There's no need for a grand protector in the forest considering the many millennia of evolved ecological stability there. Gaia is held together by an equal distribution of resources based on needs, not wants. Hobbes sees strife in nature in competition between people for the same wants, a reflection of human nature. In Chapter 21, Hobbes confesses that human liberty is an obligation to submission, presumably to the state covenant epitomized in the ruler.

The reality of the forest is that many of these differences have been resolved through adaptations. For instance, Hobbes talks about people naturally seeking glory by invading another's farm. Typically, with forests large enough, creatures of Gaia are self-sufficient. With foraging ranges depleted by invading humans, some animals raid human crops and are killed as "pests." The truth here, using Hobbes' language, is that humans are *depriving* forest organisms of their right to life, not the other way around. Hobbes, proclaiming a natural distrust among people, goes on to say how one who successfully invades another's land and takes possession will eventually be subject to a similar fate. In nature, then, Hobbes sees constant war through a fighting disposition with fear and the need for a powerful leader. He saw the human world, not nature, as it suited his political ends during a time of revolution and civil war. The opposite in nature, had he really looked, is true.

There is, putting aside instances of evolved obligate carnivory, a balance of adaptations in habitats across ecosystems. We see, rather, aggressive feeding in human hunting, poaching, and animal agriculture; most people in developed and even developing nations can choose not to raise, slaughter, and eat animals. Anyway, somewhat cynically, Hobbes believes nature sets a stage for people to destroy each other. The glory of power plays a central role in his picture of natural life: advancing it, gaining it, and holding onto it. Forest inhabitants are focused on survival and reproduction for genetic continuation, not supremacy. While there are power struggles and ranks among social creatures, from ants and wolves to primates, that prescription is not totally the remedy to their needs or desires. Had Hobbes traveled to the Americas, which he references for support of his claims, he would have found that people there were not savages without government, as he asserts, but quite sophisticated. The point is that the constant state of armed warfare Hobbes refers to is among humans, not forest fauna who evolved their own ecological niches. The problem for Hobbes' selfish humans is that they all want, need, and especially desire the same goods or territory in modern states, and thus the conflicts. In Hobbes' worldview, there should be restraints on freedoms to prevent chaos; that is not, however, the situation in the forest, and so all wildlife organisms should be left alone to thrive and service the soil, water, and atmosphere.

In Chapter 17 on government, Hobbes advocates for a strong régime without restrictions enabled to do what it must at almost any cost to maintain peace and restrain people from trying to satisfy, as he says, "natural passions" for violence (Schmitt 2009, p. 85). Rather than spreading out authority and power (his word) as we might see in the forest, Hobbes asks for a supreme ruler. This sovereign acts for everyone, but that would hardly be the case in nature where individual selection, for the most part, is dominant. Hobbes is describing a system where the ruler and small government are exempt from and above laws. This scenario is not what happens in nature where all organisms are in a struggle for existence against the pressures of the environment. In the synergies of nature, flora, fauna, and fungi are shaped by and also shape their environs. Genes respond to the environment, and this process of selection explains how Gaia is a self-regulating government, assuming no human intrusion.

From his Introduction, Hobbes says that a state is the art of humankind from nature. By the same token, there's a commonwealth of the forest created by plants and animals living together. Hobbes goes on to say that the body of state must be defended, and so too for the forest state of Gaia. Of course, Hobbes argues for a supreme leader who would presumably without self-interest protect everyone. On the other hand, natural covenants in the forest have worked wonders of non-aggression and mutual aid, generally, in the green body politic. From his Conclusion, Hobbes says human nature demonstrates a consistency for civil duties. He also talks about the liberty of people to submit with consent to a ruler, but that does not pertain to forest animals. They have not willfully submitted, evidenced by their deaths, extinctions, frustrated feeding ranges, limited habitats, and other desperations thrust upon them by humans.

John Locke (1632–1704), *The Second Essay Concerning the True Original Extent and End of Civil Government* (1690)

A liberal society equals toleration and equality in a plurality. Equality is more apparent in the wild, for organisms of all types have evolved differently precisely to fit together and function as one whole organism. Locke offers a moral theory of obligations and sociability in nature, not pure force and violence (S. Smith 2012, p. 167). The forests of Gaia, in contrast to many human nation-states, naturally promote the good of inhabitants evident from the long evolutionary history of ecosystem structures that have liberally endowed organisms with the abilities and choices to thrive. In the big picture over great time, Gaia does not interfere with but tolerates and enables adaptation and growth since all benefit from the success of each part in a diversity of life forms. There are "natural moral rights" all people should recognize, and this is different from the nastiness and brutality Hobbes sees in pre-political nature (Christman 2002, p. 43). Whereas for Hobbes without a Leviathan there's a free-for-all, Locke sees obligations toward another's property and liberty. However, it's exactly because of an understanding of natural morality that Locke presumes people will accept a civil state that enforces rights and protects possessions. For Locke, "resistance or rebellion" is justified against any unjust sovereign who counters natural rights, contrary to Hobbes, where there has been no consent (Christman 2002, p. 45). Libertarians would agree with Locke that state authority is an unfortunate necessity so long as it does not impinge on property or other rights (e.g., Nozick 1974). Though I argue for a forest state, the authority would not emerge from humans but is inherent in the rightful place of wildlife and their freedom of habitation.

Locke sees the state of nature as benevolent equality, but not unfettered freedom, where reason by each individual is employed to comprehend morality in a natural society (Clapp 1967). However, because individuals can differ to extremes and fight, a social contract enables people to form a civil government among themselves to uphold freedom, security, and property rights and to address unjust treatment. Jonathan Wolff (2006) wonders if there's any reasonable alternative to the creation of a state. If we look to other species and hunter-gatherers, we might find models. Of course, these are small populations. For Locke, individuals are naturally autonomous, but we still have the question of who governs and who submits. In a social contract, there is consent by individuals to be governed. A utilitarian might emphasize happiness for many over individual autonomy. The state is difficult to define since there are varieties, but some commonalities might include legislative power, enforcement, political power, and citizen protection (Wolff 2006). States own forests and yet don't protect them or species there but enforce unjustified violence for supposed political or national reasons. There should be some obligations of a state to its inhabitant citizens. Likewise, wildlife in the forest breaks no laws and contributes to the earth, a benefit for all.

In his *Second Treatise on Government*, Chapter 2, Locke presents, in contrast to Hobbes, a more liberal view of humans. Locke sees all humans above animals and possessing a moral standing to be respected (Schmitt 2009, p. 83). According to Locke, in nature there is equality with no one holding more power than another, unless dictated differently by God. Change Locke's word "God" to Gaia and we have elements of evolution through natural selection and descent with modification. In nature, Locke imagined natural obligations of respect for another's life or property, again according to his supreme God. In reality, self-preservation is part of the natural order in Gaia, not necessarily governed by any Godhead but through the biological maneuverings of organisms living together. Borderlines in nature make evolutionary sense; it's risky to engage in conflicts where energy can be wasted and injury is possible. Exceptions might include conflicts or displays for sexual dominance promoting gene survival through reproduction. Generally speaking, one sees more compromise in nature through the creation of ecological niches rather than in human life where individuals and nations are often in unending economic or territorial disputes in order to amass similar resources or take adjacent lands.

In a state of nature, all "men" are free to act as they will regarding themselves and their possessions within certain boundaries in this natural condition of freedom (Locke, ch. 2). In nature, all are equal and have duties for each other; power is reciprocal. Of course, Locke is thinking of "men," though he mentions "species" and says they are free unless and until one is granted a right of sovereign domination over others. This free state of liberty in nature does not enable anyone to harm himself or destroy the health or property of "mankind," Locke avers. He says "inferior" creatures are for our use (ch. 2). With metaphysical jabs, he invokes God (too often) as master of all and grants humans the power of reason to guide themselves in wise choices by not abrogating another's rights or property. Let the preservation of peace for "mankind" be the standard, he claims. Transgressors who break the rights of others under this natural law of freedom and equality should be punished. One question concerns the natural rules and borders of the forest inhabitants. Certainly, they have inalienable rights to live peacefully as they have for many millions of years, for that is the true state of nature. Locke talks of reparations owed to injured parties (ch. 2), which could now be considered for the forest state of Gaia. He adds that anyone in nature has "executive power" of natural laws. Locke tosses around the concept of and the word nature without reflecting on the innumerable animals and plants there, supposing, rather, that they belong to people through a divine creator as stated in the Book of Genesis.

Despite these natural laws, Locke sees, like Hobbes, a need for civil government consented to by all people since humans are filled with self-interested passions infringing on another's property. He predicts, ultimately, violent "disorder" arising among people in a state of nature without government (ch. 2). As far as we know, animals, unlike humans, typically do not attack another out of malicious motives. An exception might include one group of male chimpanzees, our genetically close relatives, aggressively battling another. Usually, if there's an attack among forest inhabitants, it's for defense of territory, resources, or kin. Likewise, but for some social species including ants and bees, there are no animals, like humans, who appropriate another's body or mind to work for no wages under horrible conditions. Nonhuman colony workers benefit the group and so work to their own advantage as well. A human slave works hard for another's profit with little to no gain and suffers humiliation and loss of freedoms.

As for property, Locke believes that his God has given the world of nature and its resources to "men" through Adam for their own use and convenience (ch. 5). That antediluvian thinking predominant among many people today is in blunt contrast to my argument. All beasts, or as he continues, "inferior creatures," including fruits of the land belong to humankind in common with no one having sole privileges. Locke claims that humans add value to the harvests of nature (e.g., by cooking) and those commodities thus become human property. The labor involved in plowing, sowing, and reaping makes nature individual human property, he says, and not necessarily common for all. Resources of nature labored over belong to him who works for them in self-appropriation. By this very line of reasoning forests belong to microorganisms, vegetation, plants, and animals; they are the workers of the land. There is no question whatsoever that Locke views nature as human property belonging to first stakeholders. He says whoever kills a deer owns its body because of the labor he put forth in the slaughter. Needless to say, the deer might have been a mother caring for her young while browsing seedlings and fertilizing the soil, also acts of labor. He claims the "common state" of nature supplies fauna (or beasts, to use his word) for humans to take, and whoever is the first to do so acquires a right to the meat (ch. 5). At the same time, with no irony and eerily predicting the industrial revolution and the current trend of deforestation, he says his God's earth was not made for humans to "spoil or destroy."

Locke insists one can enclose lands from the commons and claims this is an improvement for farming or hunting. He admits there should be limits to and consents by people of the commons and makes an important point when he says that enclosure need not become appropriation if there's "still enough and as good left" for others (ch. 5). By this he means for other human beings, not plants or animals. This idea, often quoted, flies in the face of what has happened since he inscribed the words. Modern humans, especially through real estate development and urban sprawl, industrial agriculture, cattle ranching, logging, and mining have taken far more lands than necessary, thus depriving numerous species from having "enough" to thrive. Deforestation is an act of enclosure since, paradoxically by elimination, it diminishes lands and entraps creatures into tiny, remaining scraps, assuming they don't die soon thereafter or become extinct. Such wholesale destruction must now be met with securing forests from humans and returning the newly enclosed lands to their rightful owners, plants and animals. Human capitalist production has provided far less for animals evident from maps in Africa, Indonesia, Europe, East Asia, the Americas, etc. It seems Locke believes his God did not intend lands to remain wild and uncultivated but rather, as he has stated and worth repeating, for the "use of the industrious and rational" (ch. 5). Enclosure is permissible with consent without prejudice to others, it seems, but in modern societies all manner of flora, fauna, and fungi have been killed or displaced.

Locke admits that human desires have changed the "intrinsic value" of nature (e.g., gold mining) based on its usefulness for people (e.g., farming grains). The value of land improves, he says, when it is cultivated for human use (ch. 5). Animals of the forest work hard to maintain an ecosystem, but that was not recognized until much later. Instead, Locke says that land left fallow to nature's desires is "waste," which contradicts the point of my argument. There are no "weeds" or "pests" in nature; nothing is wasted since forests, wetlands, and grasslands act as complex ecological basins of growth and decay that help sequester carbon dioxide, secure soils from erosion, stash massive amounts of water, and cleanse the air. Locke is looking through a theological lens only for human advantage. He insists that human labor alone puts value on land. Even as someone of his time, he could have considered the ramifications of unsustainable sourcing year after year as populations grew, noted in his comment about taking just enough. If, as he insists, labor gives the "right of property" (ch. 5) to the worker, then forests belong to the animals and plants who service Gaia. Industrialists and capitalists, as we will see with Karl Marx, believe that virtually nothing belongs to the worker except for his low paid labor. Further, the value Locke speaks of is not just people living off the land but hunting and harvesting to sell produce for a large profit. Animals, on the contrary, take only what they need and don't hoard and sell as humans do trying to gain advantages over others. Human values can be virtuous, but Locke seems to focus, like Adam Smith, on financial remuneration.

Locke spends time talking about "men" coming together in civil society to form a governmental body politic, establish laws, assign punishment, etc. (ch. 7). While this organization does not appear to occur in forests, it actually does. There are complex synchronizations from bacteria and insects in the soil to mammals and birds in trees, to waterways with fish all divided and yet acting as one body to sustain an ecosystem naturally. The real problem is that humans injure forests without penalty. The only remedy is for forest systems to be left to their own devices as a civil society without any need for judges or legal appeals, except against humans. Human predators are worse than parasites in nature. If, as Locke suggests, there is need for an arbiter in a civil society to settle disputes (not a king or monarch) to avoid anarchy (his word) and loss, why do so many creatures in nature evolve and survive? There's no chaos in nature, only an impartial randomness. In line with Lovelock (1979), the natural world is a magnificent and efficient self-regulating system of oceans and forests sustaining all life forms.

In political assembly, says Locke, the majority has power. Here he talks of the "power of the whole" (ch. 8), but only from a human perspective without looking at how well forests maintain themselves. He speaks about consent and submission, which occur in forests, too. All creatures consent to a life together and submit to restrictions regarding habitats, food sources, etc. By his own admissions, Locke says "men" cannot live together in a natural state, which ignores the mammalian and ape nature of humans, coming to be recognized at that time in the work of anatomist Edward Tyson. This negative thinking also ignores Native Americans, whom Locke refers to on occasion. Modern people are born into and under the restrictions of a government and see the world from that point of view alone. Forests work along a different trajectory with freedom from absolute or other rule but for pairings, parenting, families, groups, colonies, troops, flocks, coalitions, populations, and other ecologically friendly social structures that carry hints of moral norms. More so, various forms of altruism, such as direct, indirect, and reciprocal, demonstrate animal ethics. Not animals or Indigenous people but modern humans in the woods would subject others to their will and swallow them. Just as humans in modern societies give "tacit consent" (ch. 8) to their governments, so too animals of the forest who freely participate in and cooperate with a whole ecosystem for the welfare of all. Humans and animals alike can be and indeed are members of a society following Locke's logic.

In nature, says Locke, people are constantly exposed to invasion from others (ch. 9). That would be the human perspective. Animals and plants manage ecological niches and ecosystem territories for themselves as they participate in Gaia and have evolved defense mechanisms for protection, like changing shape or color, hiding or flying. Locke's notion of insecurity occurs in nature, as well as a need for control, but these are magnified exponentially in humans because of self-interest, evident in their population growth, sprawling suburbs, and plundering of resources for sale. The model of civil society and sustainability is in the forest. Locke says governments and societies formed in order to protect property. Indeed, that's precisely the problem, where in nature property like land, trees, plants, and water are shared. He makes many suppositions about Indigenous people, for example how they are selfish, irrational, and overly passionate about conflict. He claims these deficiencies lead people into creating a civil society, as we can see, for instance, with wolf packs, elephant herds, baboon troops, and great ape coalitions.

Continuing with his *Second Treatise*, Chapters 9 and 11, Locke diverges from Hobbes by observing how, in spite of their desires, people have a moral sense of right and wrong. Locke talks of a communal body politic where everyone is free yet observes bounds, and this picture is closer to what transpires through the mechanisms of nature that operate in communal consciousness. A big difference is that Locke sees the commonwealth as a means of preserving and protecting property. Meanwhile, in nature, all resources are shared. As with Hobbes, Locke mistakenly puts modern humans in a state of nature and so argues that self-interest would prevail and cause chaos with no resulting punishment. Anyone who looks to a forest sees that's not the case where a large assortment of species live together mostly in unison and mutual aid even if individually unconscious of the help they receive and give others. Locke says the community must create consent to laws. Indeed, through natural selection what's enabled is ecological governance, a combination of powers across species.

Locke believes the commonwealth needs legislative powers at society's consent for the public good (ch. 11). He says property is disordered in nature and why societies, boundaries, and legislators are necessary. He talks about epic power struggles and armies, which normally don't occur in the forest. Those who enjoy the benefits of government, meaning all citizens, should be happy to contribute to its longevity. Surely, animals in the forest pay through their labors and feeding to till soil, disperse seeds, fertilize the earth, etc. Their corpses become food for microorganisms that in turn add nutrients to the soil enabling more growth. Forest property is equally given and shared. Yet, Locke insists that people are ultimately the "supreme power" since they can place or remove legislators (ch. 13). While there are no legislators as a body per se in the forest, plants and animals hold power and self-representation outside of invading humans. With a government the legislature holds power; thus, without a government in the forest the flora and fauna rule as the "public will" of society. The "public weal" is tied into the fate of forest lands. Locke quotes Cicero by saying, *salus populi suprema lex esto* (the good of the public is the supreme law), which he avers should be followed as a maxim. So it is in nature, but since industrial times this principle has not been observed and hence our climate crisis.

As for prerogative or the right to act for the public good without or outside of law, discussed by Locke, forest lands provide a fine example. There are no rulers of the woodlands who hold interests separate from the "good of the community" (ch. 14) other than humans. P. J. Crabbé and J. P. Manno (2008) assert that ecological integrity is a public good. Without consent, any conquest of one nation by another makes conquerors no less than "robbers and pirates" (ch. 16), as we see with humans concerning the state of Gaia. Force upon the innocent forest state is such a crime against nature that some remedy should be put forth. If not killed by people directly or indirectly, animals become slaves to human misdeeds in a fraudulent and "unjust war" (ch. 16). Gaia has not willfully forfeited its sovereignty. Right cannot come by subjecting innocence to a bulldozer. Remedy can come by reparations including reforestation, rewilding, and setting aside great tracts to be left alone without excluding Indigenous people. Strict rules and enforcement for forest edges, wetlands, and grasslands could be granted to locals trained in sustainable farming, not to corporate profiteers. Not surprisingly, Locke sees animals as property. He says every "man" has two main rights: his own personal freedom and ancestral inheritance (ch. 16). Both of these rights should apply to forest creatures who have not willingly relinquished either.

Locke suggests a conqueror might have a "just cause" (ch. 16), but not with how humans have destroyed so much forest flora and fauna. At the same time, Locke says that those who usurp are not in the right (ch. 17). This only makes sense regarding forests if we accept that they are there for the taking and whoever does so first is the property owner to use the land as he sees fit. Life potential and growth power are in the forest, not with humans. In many ways, as suggested obliquely by Locke, governments, corporations, or nations should withdraw from forests and loosen their grip from those lands since they've acted like unruly princes and autocrats. So-called human rules of forest lands have broken a trust of nature to preserve essential resources, to wit, how the forests of Gaia from north, south, east, and west provide clean air and water for humans, too. Animals can rebel against human authority through proxies. How are these deputations to work with the very nations who do so much damage? The forest commonwealth has no self-defense against marauding humans and needs all the help it can get to prevent its demise. Locke says people, in a sense, relinquish individuality once they enter into society (ch. 19). This has been the case with forest inhabitants in their community and has reaped benefits for all, in contrast to human nation-states.

As for Locke, people can live peacefully without a government, contradicting Hobbes. Locke sees a state of nature as free and equal; freedom and equality are the laws of nature. At least in the U.S., Richard Schmitt (2009, p. 4) claims there's no "settled possession" regarding personal or corporate liberty since any freedom to act, use harmful materials, or tamper with another's rights is regularly questioned or litigated. I hope my argument proves Schmitt's point, regrettably against the natural world. Although some human rights are not settled, we should nonetheless include forest rights in that discussion. Freedom equates to debate, but so far, the flora and fauna of the forest have had little voice but for conservationists, field scientists, and animal rights activists. With a theory of forest sovereignty, balance should shift in favor to preserving forests for world health. Ronald Dworkin says the ultimate political value is equality or treating all people as equals (Kymlicka 2002, p. 3). Animals should fit into that equation. Kymlicka and Donaldson (2014) state directly that animals in the wild should be granted sovereignty over the territories they inhabit and that wild animals in cities should be granted resident rights.

As Wolff (2006) notes, for Locke, no one has a natural right to subordinate another in a feudal hierarchy with an all-powerful sovereign. At the same time, for Locke, there should be an enforcer in nature, and thus the acknowledgment of laws. This is not really so in nature among many animals and other organisms. Though a pioneer of liberal thinking, Locke's ideas ignore animal rights then and now. In terms of freedom and equality, individuality is not eliminated in the forest by virtue of natural selection. As a double-standard, Locke intimates that early humankind was permitted to take what was needed to survive. Compare those small populations with international industrial corporations. To his credit, recall that Locke says we should conserve resources and share them by not taking more than is required. Most modern societies have not followed that dictum. Against Locke's advice, we waste resources and leave nearly nothing for forest dwellers. Besides, Locke viewed "inferior creatures" like animals as common property for use by humankind (Wolff 2006, p. 139). That's unjust and not ecologically sustainable. As Andrea Wulf (2015) points out, hardly one hundred years later pioneering naturalist Alexander von Humboldt was beginning to see the wasteful destruction of deforestation in South America. If Locke believes a person's labor and work product belong to him, then what of the ants, worms, bees, beavers, birds, primates, trees, etc., and their ecosystem engineering.

The key is that Locke says one's labor mixes with the object and makes it his own (Wolff 2006, p. 140). With that reasoning humans cannot take from forest creatures what fairly belongs to them. The God myth pervading Locke is likely the reason. Equally shaky is Locke's value-added theory for ownership. Animals add value to forests by virtue of their labor with flora, fungi and other organisms in the collective bio-network. Modern humans, on the other hand, with a corporate mentality of acquisition and capitalism have devalued forests of their capacity as lungs of the earth and carbon sinks. Wolff is skeptical of Locke's arguments anyway, suggesting only temporary use and conditional possession for the fruits of the land (2006, p. 141). That dictate cannot apply to flora and fauna who live in forests. Wolff also says it's difficult to improve on Locke and find solutions to the problems his theories raise, although that's what's attempted here. Generally, when Locke suggests there's "enough and as good land left for others to take" (Wolff 2006, p. 142) he's writing before the industrial revolution. Still, this is an anthropocentric doctrine of acquisitiveness and unequal distribution that lays a path for industrialization along with colonialism.

The underlying assumption with Hobbes and Locke hints that nature alone is insufficient for humans to have fabricated large societies, cities, and especially nations. We've left nature, according to their line of thinking, but paradoxically maintain a grip exclusively for our use on any resources there. If Locke, contrary to Hobbes, can imagine a small human society in nature, then why not for groups of wildlife, too.

Jean-Jacques Rousseau (1712–1778), *The Social Contract* (1762)

Individuals must be permitted to have direct input into a social contract, essentially self-governing freedom. Rousseau finds sovereignty in the general will of people (S. Smith 2012, pp. 190, 205) or, by analogy, the "mutual freedom" of forest agents. In this collective interest individuals can also become lost. In a social contract, one's rights and freedoms must easily coexist with those of other state inhabitants. Might does not make right, according to the social contract, and individuals in a society must be free to express their rationality (Grimsley 1967). Freedom of will in society is what creates a moral being with whom others agree to associate in an open community. At the same time, while a person is good naturally, human willfulness, which can be selfish, sometimes gravitates to inequality and injustice away from a collective will for the common good. The social contract is not equivalent to a government. Rather, it depends on people recognizing the equality of each other without making burdensome demands

(Rousseau bk. 3, sec. 16). This then is the malleable and evolving law of nature where all share a common interest of sustaining the forest or what Rousseau calls in human terms intentional "civil association" (bk. 4, sec. 2). There are different centers of "political legitimacy" for these three key thinkers: for Hobbes it's in reason; for Locke it's in natural rights; for Rousseau it's in freedom (Christman 2002, p. 49). The state of nature for Locke and Rousseau, in sharp contrast to Hobbes, can be "tolerable" (Simmons 2008, p. 24).

For Rousseau, according to Ronald Grimsley (1967), inalienable sovereignty is found in and shared with the authority of a collective. Individuals must actively participate in society where justice is not divided from utility. By now, readers can see the analogy with wildlife. Rousseau believes "social order" is a right (bk. 1, sec. 1), just as we see among forest inhabitants. However, he does not view the right to social order in nature itself but through human conventions, following Locke. He asserts that the domestic unit is the original template for "political societies" (bk. 1, sec. 2), a family model in varying forms found in nature. Rousseau talks about the weakness of those believing they are strong who fail in virtues of right and duty. Forest flora and fauna cannot resist human forces that enter their land and, as he suggests, such force is not moral (bk. 1, sec. 3).

Rousseau has famously said that people are born in freedom but ultimately chained by circumstances. He talks about how people and slaves of kings feed the royal household but could in mass object to enforced bondage (bk. 1, sec. 4). Forests feed humans but don't have the ability to rebel on their own. The thought Rousseau uses for the state of slavery is alienation on some physical level, but those in bondage can be free in other ways. Renunciation of liberty is not human, he states bluntly, but humans have by the early twenty-first century taken over what remains of most forests. Generally, no organism would relinquish its freedom or sense of control, either, according to the laws of biology.

Humans have enslaved forest lands as if they don't belong to the flora, fauna, and fungi there. Humans attack forests as if in war and act as ancient victors vanquishing all there with a false sense that they have a right to enslave or kill the conquered. This primitive attitude is described in ancient writers and should not be expected today. As Rousseau knows, war against nature is about selling things: acreage, lumber, minerals, cattle, etc. Within nature there is no "constant property" and so no war, he observes (bk. 1, sec. 4). That stance was before the industrial revolution and corporate agriculture. At the same time, Rousseau says there's no "social state" in nature (bk. 1, sec. 4), but we've learned through field scientists over the past one hundred years since Darwin that there are laws of nature regarding growth, life, death, reproduction, possession, eco-habitats, etc. There's "normative regularity" of social conformity among animals (Westra, et al. 2024). Rousseau proves the point when he says that war is between nation-states with individuals suffering the consequences of belligerent rulers. Since no single entity rules the forest because it's a self-governing collection of ecosystems, flora and fauna cannot win a war waged by humans. Corporations and others enabled by any state to ravage a forest is a mercenary act of politics, an extension of political greed, and not socially moral.

Forest inhabitants are under no obligation to humans but freely service and cleanse the atmosphere. As Rousseau suggests, humans subdue but do not rule nature (bk. 1, sec. 5). Maybe with wise guidance tempered by restraint, forests can again be free. The human body politic has no interest in the public good if it continues to deforest lands for development, logging, farming, mining, or cattle ranching. Forests are measurable public goods (Reid & Lovejoy 2022, p. 169) if one uses data about carbon capture, atmospheric and soil hydration, biodiversity, etc., all of which are more crucial to world climate health than GDP. Section 6 in Book 1 of Rousseau's text is called "the social compact"—a "sum of forces" coming together to keep individuals free while maintaining order and self-preservation. Before modern humans invaded forests this was so. In their mutual alienation, forest flora, fauna, and fungi associate to increase a common good and healthfulness. Each organism sponsors all in a roundabout way. We see a stark difference in most human societies where land, power, and wealth are held by some at the top of a hierarchy. The social compact, according to Rousseau, requires individuals to join in contributing to a general will (bk. 1, sec. 6). In fabricated (human) societies, many always at peril for civil war or political collapse, we find what Nietzsche might call a will to power. The common will of equal consideration, on the other hand, appears more uniformly in nature.

Since Rousseau emphasizes education, we should consider what type of education we have provided our children. Teaching them about kindness to the earth and animals tied to ecological practices is paramount. Lots of education emphasizes materialism, consumerism, and earning potential, especially on the college level. The mother-infant dyad in nature is one of education: where to go, what to eat, how to survive socially, etc. As Rousseau might say, this is service and training "not contrary to the will of" the forest state (Wolff 2006, p. 78). In spite of acting with self-interest, wildlife produces what is good in a balance for all. We don't have to prove there's a willful, conscious intent of organisms in the forest to generate goodwill. It happens as we can see from the historical and geographical net result. Forests survive until touched by the modern human hand, also obvious. That animals create a common good is akin to Rousseau's social contract equally enforcing everyone's interests. The forest law works since there is no central power, no class structure, and no corruption across species. Within some species there is status and hierarchy, but even so the animals are aware of how equally they are affected by changes in resources or forced migration. Individuality is not eliminated. The law of the forest governed by Gaia almost equally affects all inhabitants, with exceptions for ecological niches. A species group might have a particular interest, but all species together hold an unconscious common interest of forest survival.

A large issue is that there's no social contract stating explicitly that human political states should let forests alone. That's odd considering how grasslands, wetlands, and woodlands uninhabited by humans or their farm animals act as carbon sinks to clean air and water so that people can avoid social, food supply, and climate catastrophes. Rousseau claims a civil state exists only outside of nature since the wilderness lacks justice or morality (bk. 1, sec. 8). This is not so, evidenced by the stable state of affairs in the forest and explained, to name one instance, by authors like Marc Bekoff and Jessica Pierce in their book *Wild Justice* (2009). Rousseau seems to denigrate physical impulses and prefers duty, but the evolved instincts and social collectivity of forest creatures has allowed them to adapt ably and maintain their Eden for millennia. As with Locke and Hobbes, for Rousseau the civil state means people can have property ownership. That limited view is the real problem since it extends to deforestation for personal or corporate gain.

Rousseau talks about the "right of the first occupier" (bk. 1, sec. 9) where the primary condition is uninhabited land. For many millennia land has been managed by all manner of species from microorganisms and mammals to ungulates, birds, and trees. Rousseau goes on to echo Locke about how through cultivation land belongs to the farmer because of his labors. That too applies to the animals, from worms to beavers, since they are the first and foremost ecosystem engineers in place before human settlements. To his credit, recalling Locke, Rousseau further suggests that an occupier should take only what he needs, leaving the remainder for others. We know that this is not the case when multiplied by so many people, nations, and corporations over hundreds of years. Part of the problem stretches back to the ages of conquest and then the era of colonialism. Rousseau goes on to say of kings and other rulers that they don't just possess the land but hold the inhabitants, alienating them from secure ground. The same can be said of how modern humans have treated forests in illegitimate possession and the shrinkage of habitable space.

The social contract establishes a moral equivalency among people, so why not between people and forest creatures and plants. The more we learn about the intelligent behaviors of animals, from their cognition and intelligence to socialization, tool use, and multi-modal means of communication, the more we should consider them as our biotic equals. Rousseau offers the idea, in line with what's proposed, that the sovereign is a collection of beings (bk. 2, sec. 1). Later, he intimates how the state is moral through the care, unity, and preservation of its inhabitants. This observation is true in some countries more than others. At the same time, everyone who is part of a community and consents to a social contract makes sacrifices as well as contributions. I've routinely enumerated the contributions of forests to earth's health. As for sacrifices, animal and plant inhabitants, while free in the forest, cede space to others, do not have exclusive rights to all resources, and at times succumb to others where in death their bone marrow feeds the soil. Loaning words of Rousseau, the animals work for many while laboring for themselves (bk. 2, sec. 4). Every living organism, whether human or not, wills self-preservation and control with the understanding that risks are involved.

In a human state the social contract gives life to the body politic (Rousseau, bk. 2, sec. 6). In the forest, the flora, fauna, and fungi taken together create Gaia not by legislation but by existence itself. Laws need not be written. For example, there are laws of physics and gravity. Rousseau quotes Montesquieu (bk. 2, sec. 7), saying that institutions shape a ruler through laws established by the consent of legislators. In nature this occurs by analogy via descent with modification; with the engine of evolution as ruler adaptable species are the legislators. Humans seem determined to impose their legal ideas on nature when Rousseau explains, quite to the contrary, that different places hold diverse customs and should not be expected to comply and become swallowed by burdens imposed upon them by an outsider. Politically, people constitute a state all "sustained" by its territories in proportion (bk. 2, sec. 10), precisely the same for wild animals.

At this point in his discussion, Rousseau says that legislation should not only address what is seen but more importantly what is foreseen (bk. 2, sec. 10). While some of that foresight peeks through United Nations declarations, environmental reports, country pronouncements, and executive orders, there is not enough sweeping vision for the complete reforestation, protection, and conservation of the global state of Gaia. In speaking about any questions concerning the notion of good (bk. 2, sec. 11), Rousseau says liberty and equality are necessary. Liberty, he says, flows from equality. By equality he does not mean exact parallels but through degree and rank, which translates to species differentiation, niche construction, and habitat ranges in the forest. Species evolved to tolerate each other and often feed on different parts of plant or animal resources permitting an equal distribution. With human intrusion, this energy balance is sent off kilter and encourages violence from privation while impoverishing forest creatures. No virtuous state would act similarly to itself, so why to Gaia. Each organism in the forest, from the tiniest insect to an elephant, contributes a share to the sustainable whole. All flora and fauna influence forest growth. Relations might not be equal in scale, but the results of relationships establish environmental uniformity.

There's an invisible and symmetrical geometric contour to the forest not seen in human states. Rousseau admits that no one type of government fits all since there are so many sizes (bk. 3, sec. 1). He speaks in terms of the shapes and proportions between people and sovereign. The virtue of Gaia is that no single entity rules but that all dimensions are proportionally sized and fit together in unity, which is more than can be said for many human nations unraveled by all sorts of political, cultural, and social discord. Moreover, as Rousseau suggests, the proportions of the state are not reducible to the number of bodies but to their actions and consequences (bk. 3, sec. 1). This thought, worth emphasizing, is how the state of Gaia consists of a self-regulating arrangement of plant and animal ecosystem engineers, each playing a role for symbiotic health. The self-governing system of ecological managers possesses, to use Rousseau's words, "sensibility common to its members" for the preservation of all (bk. 3, sec. 1).

There is danger, Rousseau says, when a ruler's self-interest contaminates the affairs of a state. While this happens among humans with greed for money or power, self-interest in the forest is balanced with sustaining the whole, clear from its successful evolution. Rousseau claims a real democracy is against "natural order" since the populace cannot really govern the state of affairs on a daily basis (bk. 3, sec. 4). Yet, that is what occurs in the forest as microhabitats operate in conjunction with so many others to form a grand collection of well-functioning and interconnected ecosystems. In the forest this form of management succeeds for the very reason it sometimes fails with humans: constant shifts and changes. These altering forms require maintenance, but that's the law of evolution and adaptation where organisms struggle to fit into a changeable environment. The more pertinent question concerns how so many humans and self-interested nation-states will adapt cooperatively to the existing climate crisis. Worse, some politicians, corporations, and nations resist fighting climate change since they benefit from the polluting technology and related industries. People want to consolidate power and pass it down to family, friends, or business partners. Forest flora, fauna, and fungi share heritable energy, not power. Any president, prime minister, or monarch might know the people he or she governs, but such a person cannot grasp the needs of forest inhabitants. These rulers pursue the interest of the party council, business leaders, or constituents and not interests of the forest.

Rousseau cannot state the best form of government since so many nations differ in needs, populations, climates, etc. This might explain why there's no need for a formal, internal forest rule when there is, rather, a collective equilibrium of individual wills. According to Rousseau, a government dissolves when it "passes from the many to the few" (bk. 3, sec. 10), which happens when humans overtake wooded lands. No social contract of natural liberties is broken within a forest or among its creatures. However, with humans, as he says, the dissolution of a state results in anarchy, but that's not the case in a forest since the parts hold together unless split by rare unnatural or probable human forces. There are no tyrants or despots in a forest. Rousseau rhetorically asks that if an empire such as that developed by the Romans succumbed to dissolution, what then could endure. Forests, including wetlands and grasslands, with their natural systems joined as one Gaia are the answer to longevity, not any short-term human administration. The laws of nature are the executive powers pulling together the sovereignty of Gaia.

Since the earth's forests provide clean air, water filtration, carbon sinks, rainwater catches, etc., as life givers they and not people should be governors of the planet. Modern humans are, essentially, foreigners to the social compact of the forest because they have not consented to but break the contract. Rather than a state of conflict, the natural state is one of mutualism. In nature, accounting for struggle and privation, organisms in any ecosystem work and live close together. A hungry tiger has little choice as an obligate carnivore. Even so, his feeding behavior creates forest population equilibrium and symbiosis. What he does not consume is nourishment for birds and small mammals, microorganisms, and bacteria. Eventually, the bones of the carcass become minerals in the soil. Overall, for Rousseau, it's not only self-preservation but life among other beings, examples of which we see in nature. Rousseau paves the way for the sympathetic moral sensations of David Hume and Adam Smith, but such feelings exist in nature from the mother-infant dyad and among kin. Rousseau mistakenly says that animal "brutes" have no free will or ability to improve themselves (Wolff 2006, p. 27). Many animals make choices about whom to befriend and where to go; they take interest in the activities of others so as to learn new techniques for food processing, etc. These activities sustain the forest.

Adam Smith (1723–1790), *The Wealth of Nations* (1776)

The wealth of a country is not gold stored within its borders but depends on the amount of consumable goods it manufactures and distributes (Sprague 1967). In his book, *Theory of Moral Sentiments* (1759), Adam Smith relies heavily on sympathy and fellow-feeling; in *The Wealth of Nations*, self-interest to command capital reigns. In *Moral Sentiments*, Smith says humans are self-interested but have the ability to become impartial observers and form a sympathetic perspective from moral judgments. We see Smith juggling contradictions. The capitalist, guided by an invisible hand (Smith's phrase), contributes to the welfare of others. As a moral philosopher, he's concerned with how individual self-interest can benefit society for the common good. Nevertheless, Smith's structure takes the individual out of a community and into a factory contrary to the ecological economics in forest systems (Costanza, et al. 1997).

In the eighteenth century, there were very harsh working conditions for those in mines, on farms, and in factories: long days, dangerous work, and low wages. From his travels in France and discussions with Dr. M. Quesnay, says Robert Heilbroner (1961, p. 34), Smith came to realize that wealth is not cold cash but production and exchange circulating via *laissez faire* policies. Smith germinated this idea, emphasizing industrial manufacture to help fuel the engines of empire. As Heilbroner points out, Smith appears as an advocate of the working class and not the rising bourgeoisie whose morals might be doubtful, but his real focus is on national wealth. For Smith, wealth is not held by any one class, much less royalty or nobility, but moves freely among all consumers. Not dismissing self-interest, Smith ultimately finds market concerns contributing to the shape-shifting body of a social whole. Self-interest and competition lead to the production of materials that ideally satisfy needs and desires of the populace, suggests Smith (Heilbroner 1961, p. 40).

Consider the previous paragraph in terms of Darwin's natural selection. Primarily, the aim is financial pay and perhaps wealth accumulation, not charity or mutual aid. Most humans across continents have very similar needs for the same resources, in contrast to other species who have adapted to different habitats or have hybridized over time into another species with a new ecological niche. In the human realm there is total competition among producers and sellers with competitive prices for products. As we saw in Peter Wohlleben (2016) in Chapter 1, echoing Darwin and Kropotkin, while there is competition in the struggle for existence, many species unconsciously engage in mutually beneficial actions. Competition regulates society, an economist might say, just as a forest biome is self-regulating. Totaling the net result, we can see that the human version of self-regulation is harmful to the natural world, to say nothing of the crime and punishment in the capitalistic social sphere. Ultimately, evaluate provisioning. While some individuals and species die in the wild, overall, the effects of Gaia are more advantageous to all than those seen in the human realm, as any examination of history can show.

Indeed, Heilbroner (1961) points out that Smith acknowledges a wealth system of self-regulation, with the caveat that market forces can be unforgiving for those at the top as well as those at the bottom of any social scale. Today, in contrast to Smith's era, there are highly sophisticated tariffs, reams of government regulations, labor unions, complex banking transactions, and other capital or industry manipulations at work. In Smith's time, factories were able to employ many people and produce more goods cheaply for a large population. Children and others were employed with a wage close to the poverty line. The forest for some corporations and nations is like a factory, where the aim is to divide and simplify labor. The difference is that the forest is impoverished by these human actions. Like early eighteenth-century industrialists reacting to market demands by making and accumulating wealth, the same is happening to forests with little regard for the lives there. Smith agreed to the accumulation of wealth, calculating how investments in machines meant division of labor creating social welfare. This equation bedevils us today where improvement and progress along with jobs and wages are links in an endless chain reaction. Just as Smith says in terms of consumers, let the market alone, so do Wohlleben and E. O. Wilson regarding the forest state of Gaia. Smith, contrary to how industrial capitalists used *The Wealth of Nations*, was concerned about the health and happiness of workers and did not wish to see them swallowed by unforgiving machines. We should have the same ethic concerning the flora and fauna in forests. Smith is not a conservative economist in spite of his avowal of a free-market system; he seeks to increase social welfare. As a philosopher, Smith questions the motives of capitalists, says Heilbroner, wary of profit accumulation without regard for workers and the benefit of consumers.

Of *The Wealth of Nations*, Stefan Collini (1989, p. vii) says the work praises the advantages of leaving individuals to their own economic devices. Smith notes how the division of labor increases productivity. We see this in forests, where the work of microorganisms differs from that of tree-dwelling primates. Apes eat fruit and spread seeds, and their droppings become fodder for microbes who break down sediments to enrich the soil. This sustainable and organic industry of greening earth cannot be duplicated by humans who mostly rely on machines and chemicals to farm food, much of which is fed to animals slaughtered for more human food. Contrary to what Smith repeatedly highlights, unless disrupted forest production of great flora (as massive trees) and small (as mosses) is routinely completed in natural cycles. Human machines are the enemy of the forest and eradicate what took decades to green. Machines, Smith goes on (bk. 1, ch. 1), were developed by those in the division of labor who wanted to facilitate work more easily and efficiently. These noisy and dirty machines wasting water, spewing fumes, gases, and oils, have not been an advantage to forest life, which supports global health. Smith points out how not only the division of labor for one product maximizes productivity, but also how other industries whose labor is equally divided contribute to a whole network of suppliers, manufacturers, shippers, consumers, etc. All of this commerce has evolved naturally in the forest without the use of money, bills of lading, bank notes, or loans and where all inhabitants profit from each other's labors. A worm works with microbes to tend the soil, helping plants grow, which become food for mammals, and so forth.

Smith attributes the division of labor to human nature (bk. 1, ch. 2), but this is false when looking at, for instance, social insects. Smith talks of *quid pro quo*, barter, and treaties in the purchase of labor, services, and goods. He sees the machinery of finance and commerce as a boon, which added up is true for humans. In nature, while there is reciprocity and mutualism, and while in some social animals like primates there can be exchanges of food with the expectation of a return favor, most actions occur routinely as part of a self-regulating system of silent and almost invisible exchange. Contrary to my whole argument and against science, Smith claims that animals are of no use to each other, though he weakly uses as an example domestically bred dogs (bk. 1, ch. 2). For Smith, the division of labor is limited by the extent of the market (bk. 1, ch. 3). This makes sense in human terms but not on the forest scale where the "market" is always large enough so that the machinations of the tiniest creatures produce a measurable effect, unless upset by an outside force.

Early human societies bartered cattle as money. However, when there is an excess of such goods, the superfluity poses exchange problems, which were alleviated by coins to parcel out needs. We have no such commercialization or commodification of resources in the forest. Water, for example, is freely available to all without transferring coins. We know that the labor and productivity of forest flora and fauna add to the value of those lands because they are logged and mined by humans for marketable resources. As Smith might say (bk. 1, ch. 4), forests have value-in-use for the animals living there and by consequence the atmospheric moisture and cleansing powers of these vast woodlands. Humans seem to ignore this worth even though they benefit from how forests traditionally control, like a thermostat, the earth's climate. Rather, humans look to value-in-exchange, or how they can strip a forest of wood, minerals, metals, etc., for sale. In the balance, value in the labor of forest animals environmentally outweighs in its proven sustainability any deforestation activity by humans.

The object of nature's evolved political economy, in contrast to that fabricated by humans, is not gold or silver. The political economy of Gaia is production and use for the inhabitants in a reciprocal cycle of growth, decay, and regrowth. Money does not exactly circulate only in one nation but among nations; even so, there are financial crises. This artificial flow is not the case in nature, which is remarkably self-sustaining with exchange and transfer of life forms across species in various compounds of resources such as food or shelter. The effort of nature's political economy is not, as might be the case with some human states, accumulation of wealth but the fair distribution of sustainable assets. Barring a natural disaster, there is no unfavorable balance of trade, except for human interference. Just as Smith would condone *laissez faire* and free markets for the enlargement of capital and the wealth of nations, so too might humans adopt a non-interference stance toward forests so that they can flourish abundantly and provide global climate health. In other words, it's good business for the sake of a nation's security to cease deforestation and begin replenishing those lands with native flora and fauna.

Smith says a common complaint among nations is the scarcity of money (bk. 4, ch. 1), truly not an issue for forests but yet why they are ravaged so as to balloon the coffers of states, often at the expense of local people on the margins or Indigenous people in forests. The value of money is really in its purchasing power, so through the complex, modern banking system of loans, interest rates, and credits, corporations are enabled to work on forest lands with borrowed money encouraging rapid productivity to increase capital and generate newer projects. Smith says it would be absurd for a country to have more pots and pans than they need for cooking, and so too in the forest. There are enough resources to go around for all animals. There are checks and balances where flora and fauna decay and die to give light and provide space for renewable growth. In human terms, cities just get bigger and more populated as they sprawl infectiously ever outward into nature's areas. In one respect, cities are a solution to deforestation, but they need to maximize space, utilize walking and bicycle paths, increase green spaces for pollinating insects, birds, and small wildlife, recycle water and energy, etc. Intensive veggie farming with water capture and reuse in urban areas, whether through malls, rooftops, abandoned parking garages or buildings, ruined lots, unused airfields, barges, schoolyards and vacant athletic fields can mitigate marginal farming near forests to let those areas increase in size by rewilding.

When capitalists look at a forest, they do not see a cooperative whole sustaining itself but abundance for their taking, to reap products that fill the desires of consumers whom they've solicited with shrewd advertising campaigns. Consumable commodities, like those from forests, need constantly to be replenished for human wants at a cost to the flora, fauna, and fungi who produce them. In turn, these consumables when sold are turned into more durable commodities for nations and corporations in the form of cash. Sadly, this money is then used by some nations to engage in wars, which, in addition to devastating people and towns, also ruin areas of natural growth and wildlife. Relatedly, many consumer products manufactured by industries are quickly discarded as waste, perpetuating a negative environmental spiral. As with Marx, we see in the current human political economy forced labor. Borrowing from Smith (bk. 4, ch. 1), human nations extract as much as they can from woodlands so as to import the resources to others at a cost. Profits are made all around, but none are shared with the exploited animals who are left in diminished habitats.

John Stuart Mill (1806–1873), *On Liberty* (1859)

John Stuart Mill emphasizes freedom and happiness through toleration and progressive thinking to enlarge the common good (Schneewind 1967). When government should step back and permit personal freedoms, we have liberty, a modern idea. David Miller (2003, p. 65) says any "self-regarding" act of thought or expression, an idea from Mill, is acceptable and should be permitted if it does not harm the actor or another. Mill defends self-regarding freedoms against state control and why a forest state of Gaia should not be ruled by humans. Indeed, in 1948 the United Nations issued a universal declaration of human rights focused on freedoms like movement, expression and communication, relationships, proper sustenance, etc. We can see those rights belonging to forests but limited by the control of human governments and their economies. The unethical version of political economy that ravishes nature for human consumption is not a sustainable business for any life forms and definitely not for climate health. Though initially open to a *laissez faire* economic theory, Mill eventually expressed concern for social and not necessarily capitalistic goals, the latter of which he feared could be used to suppress liberal freedoms. Mill recognizes, as we see organically in nature, where genes respond to environment, that human character can be positively influenced by a morally educated society. This last thought is important in the argument for cultural evolution across human populations that will embrace steps necessary for forest sovereignty. Like Smith, Mill is concerned about individual social responsibility and the common good. He favors, in line with Smith, market competition but without unchecked growth or the absence of waste recycling, self-regarding checks we find in nature (Costanza, et al. 1997).

As Robert Heilbroner (1961) says, one of Mill's prescient contributions to political economy is his idea that production, not distribution, reigns supreme. Production is to nature whereas distribution (e.g., of wealth) is to social norms. This means, according to Mill, that economics and how a society pays workers, taxes income, and shares profits is seated in an ethical realm of progressive change. It seems, for Mill, that liberty stems from utility and not natural rights (Christman 2002, p. 108). This ethos might now focus on the preservation and not destruction of forest life because of the free-market system in capitalistic societies or those developing nations seeking wealth accumulation. There can be creative ways for states to preserve forests and yet "capitalize" from their riches by limiting farming, development, mining, and wood production to margins while leaving a vast core alone.

In his Introduction to Mill's *On Liberty*, Stefan Collini (1989) says the philosopher focuses on the worth of an individual. Mill prefers active and energetic individuals without a blindly conformist society, so something more along the lines of liberalism and not socialism. At the same time, not unlike the radical Marx, Mill looks to socialism as a means of tempering capitalism. Nor does Mill exactly follow Smith's *laissez faire* dictum, either. Instead, he imagines the happiness of others in a Victorian era of ethical hard work, evangelical earnestness, and physical self-restraint, the last two not as ingredients of free-market capitalism. Mill views diversity of character and liberal thinking as essential to a society and crucial for progress. Mill seems to distinguish between moral and sociological truths, where the former concerns individual liberty while the latter emphasizes the effect individual liberties have on society. One cannot morally force a person to intervene in another's welfare just as one cannot be forced to hurt someone. Interference is justified, Mill would argue, to prevent harm to others.

Consider how these ideas, especially harms to animals and subsequently to forest eco-engineers, align with an argument about our climate crisis and Gaia. The harm is not difficult to see, measure, or debate. Mill would probably not advocate the abuse of power we see by some nations regarding their forest lands. This means that Mill, as a utilitarian, is not really in favor of natural rights but leans more toward legal rights as he prefers representative government. Collini (1989) says there's an implied question in Mill that touches on a claim of cultural evolution awakening people to the value of forest wildlife. That is, can a society develop morally? People can evolve culturally to value nature for its own sake, but that change hinges on numerous participants, from news media and social media to educators and politicians.

Mill says *On Liberty* (ch. 1, p. 9) is about civil or social independence and the type of societal control or power that should not infringe on any individual. Yet, individual liberty and pleasure can depend on how the actions of others are limited. Morality and norms could derive from the interests of an ascendant bourgeoisie (ch. 1, p. 10). Like Locke in his *Letter Concerning Toleration* (1689), Mill is focused on religion and considers intolerance as part of human nature. Any such "religion" can be social, and those norms can prevail against animals. Mill's main idea is that people will interfere with another's actions for self-preservation, but power over any other can only be justified if it prevents harm. For tolerance, he goes on to note how there are members of society who deserve care and who should be protected from injuries, like children, the infirm, or the elderly. Mill does not seem to include in his analysis what he calls "backward states" (ch. 1, p. 13), by which he means the dark corners of society but also people fated to follow a despot. He says those who cause injury, even in cases of neglect, should be held accountable (ch. 1, p. 14). Readers should see the obvious connections here to the argument about tolerance of and avoiding harms to forest organisms.

Mill is aware that one's actions can have ripple effects. He asserts that true freedom means the pursuit of individual needs without meddling with others who act likewise (ch. 1, p. 16). We see this way of life among forest organisms who evolved adaptations mostly to ensure their own and others' freedoms. Alternatively, humans tend to pry into another's affairs, sometimes with ill effects. Like Locke, Mill preaches a code of tolerance explicitly stating that everyone gains in a free society assuming there's no harm to others and where one is not forced into supporting another's interests. As a utilitarian, Mill favors what is good for people overall. Let's include in that belief the welfare of Gaia since freedom and health there are directly connected to our own. Opinions contrary to what's espoused here are not merely, as Mill might say, a "private inquiry" (ch. 2, p. 20) but can have dire effects when any powerful politician shouts that climate change is a hoax. Readers can see from good evidence and reasoned ideas that arguments for forest sovereignty are based in widely accepted scientific facts and ethical norms. The capitalist's uninformed opinion that forests should be cleared to make money has been tested and, like trickle-down economics, has failed. People are free to have and express their beliefs, but there are many wrong and pernicious opinions accepted by unthinking individuals prejudiced by influential politicians and business leaders.

Mill suggests (ch. 23) that open-mindedness allows for the holding of contrary opinions so as to arrive at a truth. Good judgment is the capital of a thinking person who weighs facts, history, and philosophical ideas. Challenges to truth are indeed necessary, or what is often called the scientific method of testing hypotheses. Mill notes (ch. 2, p. 25), and this is the point, that some beliefs though untrue are held sacred for the supposed benefit of society, like corporate meat agriculture or oil drilling. What I'm opining will be viewed variously by an environmentalist, a philosopher, a theologian, a cattle rancher, a politician, a corporate executive, etc. There are, however, some basic truths related to establishing forest sovereignty on which many others will agree as elucidated here. Mill writes during a time when many men of the working classes and almost all women had no voice, but that's not the state of affairs today. Nor is the lack of education an excuse for not having an informed opinion since there is so much reliable information, peer-reviewed papers, etc., freely available on the internet and in public libraries.

Compounding Mill's Victorian temperament is the prevailing European inclination toward colonial empire that explicitly set out to conquer other lands, including forests. In Chapter 2, Mill spends much time talking about the test and endurance of truths. He suggests that reticence regarding intellectual ideas is a lack of moral courage and a loss to humanity. Of course, an opinion must be questioned and not blindly accepted, backed by time-tested philosophy or contemporary science. Some truths are not realized until they are experienced, says Mill (ch. 2, p. 44), which is what occurs nowadays with food crises, flooding, droughts, and species extinction stemming from the climate crisis. All of this real experience adds validity to the argument for setting aside much forest. As Mill says (ch. 2, p. 45), truth should be taken from reason, not unbridled authority. Learning should arrive through disputation and analysis of multiple ideas, the scientific method. I've heard and read the arguments for how happiness is a warm planet (Moore 1997), but that false belief does not hold water. Mill is clear that an individual's opinions and acts should be allowed free rein and expression if they do not harm others (ch. 3, p. 56).

This pronouncement of individuality is confusing because it leaves open a blurry line of demarcation and, worse, does not include harms to nature. If, as he goes on, the assertion of individuality is essential, we'd have to include nonhuman forms with whom we intimately share this planet and whose individuality is at least in part responsible for our vitality and longevity. *On Liberty* was published the same year as Darwin's *Origin of Species*, and it's not as if Darwin's book came out of a vacuum. There were others writing about natural subjects, geology, and evolution (Eiseley 1961). Granted, Mill's focus and purpose is on humanity, the crux of the problem. While some will criticize me for saying this, he neglects animal rights at a time in England when there were rumblings from vegetarian societies and anti-vivisectionists. Mill says choice is more important than custom (ch. 3, p. 59), and certainly he was on the forefront with writings against the subjugation of women and for socialism. So, it only stands to reason that I invoke him in the name of forest sovereignty. It's hard to see this, though, when in Chapter 3 he goes on about the great works of "man" in building, farming, and even warfare, where he places the good of humanity at the center of the world evident by capabilities of industrial dominance. I'll have more to say about human exceptionalism later.

Maybe it's unfair to include Mill in this discussion since he bemoans, to some degree, the collective conformity of groups that have, like a city crowd, swallowed the exceptional individual. In the forest collective groups work well, but each group consists of individuals, all of whom service aspects of the ecosystem they inhabit. The forest is not merely assimilation but an explosion of species diversity, what Mill could praise. In observing liberty, Mill asks what life means for an individual and society (ch. 4, page 75). I pose a similar question in highlighting the evolved adaptations of forest flora and fauna for survival and reproduction and yet their uncanny interlocking associations. Moreover, forest dwellers and trees not only provide sustenance and protection for each other but also for life forms outside of the forest, like marine life and human beings who rely on cool, clean air currents.

While, as Mill suggests, there are individual interests that are far different from those in society, that's a human perspective and neglects the interspecies cooperation among plants and animals. He sees no real good in an obligation for a social contract (ch. 4, p. 75), contradicting Rousseau, but admits that since all people benefit from society, they should make a return contribution. For example, he says injury to another should be avoided and sacrifices should be made in defense of self and others. We see those norms in the forest, too. As a radical idea, forest animals perceive and respond to their environs as kin (Tague 2020). Modern humans don't distinguish wild plants and animals in that way. Mill proposes total freedom for one to act and, depending on the action, suffer the consequences. That, too, occurs in the forest. He sees society benefiting from the freedom of all to self-fulfillment, and that's what happens in a forest. Meantime he cautions against self-indulgence in "animal pleasures" (ch. 4, p. 78), yet animals service the climate on which we all thrive. If utilitarianism is for the good of many without harm to others, then why not include plants and animals in that equation.

Encroachments on and injuries to another are not part of freedom and should be met with, Mill says, moral reprobation and retribution. Moral outrage is the very case to make for the human ills thrust upon the state of Gaia. Mill lists a number of harms, but two that stand out and apply here are severe dominance and unabated engrossment (ch. 4, p. 79). Liberty does not mean harm or damage to others, yet think about all of the animals caged in zoos, aquariums, medical and university laboratories, pet stores, factory farms, etc. Since people generally do not want to have their freedoms controlled, we can say the same of animals, including forest creatures. Mill spends some time talking about free trade and how there should not be any restraints since, as a "social act," trade is a liberty (ch. 5, p. 95). There is no legitimate control of the forests, particularly for trade, if it creates a public harm not only to those animals but to the earth's climate. Sure, nations have protected lands, preserves, and parks, but in some cases in the U.S. ranchers are allowed to use those lands and kill any wild animals they perceive as "predators" to their valuable cattle eventually sold as meat.

If states and governments stand for people, then forests should be sheltered from humans and preserved for global health. Ravishing forests benefits a handful of people and harms nature. Politicians reap favor, corporations and other businesses enrich themselves, and the products of logging, mining, corporate agriculture, cattle ranching, housing development, etc., enhance those businesses. It's hard to believe that Adam Smith or Mill would have imagined or condoned the scale of this destruction. Subsequently, the resources stolen from the inhabitants of Gaia are manufactured for and purchased by consumers. Those acts and consequences do not justify deforestation, which has negative health impacts on many people, by diminishing and extinguishing biodiversity important for a healthful planetary symbiosis. Animal lives and habitats as well as climate stability and forest sustainability are exchanged for money siphoned to the rich and powerful. Contrary to the business perspective, forests are not a commodity but the life blood of earth's soils, waters, and air currents. Interference with the forests is not for the sake of others but actually contrary to liberty in how climate change through heat waves, mud slides, floods, desertification, etc., are limiting people's freedoms and imposing cruelty imposed on wildlife.

In his life of Mill, Michael St. John Packe (1970) says *On Liberty* is fundamentally an essay about toleration. We should apply the same logic of this liberal ethic to forest flora and fauna granting them sovereignty to remain and thrive in their protected lands. Packe goes on to say that Mill writes on freedom of the individual at a time when emphasis was placed on a group (e.g., the government) or the people (e.g., socialism). Mill's writing is more about *freedom to* rather than *freedom from*, says Packe. Mill is less to the collective, but as in a forest, bee hive, or ant colony, or better yet a wolf pack, individuals create the whole. Unaware of cultural evolution, Packe says that "Society is not an organism" (1970, p. 405) so that individuals can function in complete freedom. People rich and poor are bound by all types of circumstances. This is true even in the forest with the difference that those individuals form an organic wholeness that benefits all beings, both in and out of forest environs.

On Liberty interrogates the limits of government interference in one's life. Mill's answer, according to Schmitt (2009, p. 146), is when protection of others over self-interest is necessary. That's not what commonly happens from the Americas and Europe to Asia and the tropics, which permit logging, mining, and farming. According to Jonathan Wolff (2006, p. 105), Mill would grant more power to individuals than to the state. Local people in the Congo River Basin are stripping forests of wood so they can cook. In other areas of Africa people kill great apes for bushmeat. Mill would ask about the conflict between individual and social rights. Forest preservation, chiefly on the perimeters where encroachment begins, leads to equitable benefits for all if required by governments. As Mill says, if someone acts with prejudice against another's interests, the government should intervene. First, this account probably did not consider forest plants and animals. Second, this policy assumes the government will have and adhere to a code of ethics regarding forests and Indigenous privileges. Sustainable management should not simply exist on a document in a politician's drawer but be alive on a daily basis.

Humans weaken forest boundaries and health. First, humans act as if they alone have complete freedom to do as they wish regardless of inherent rights and interests of other species. Humans tend to abuse liberty at the expense of others. Second, in the forest humans wield absolute control over the lives and freedom of those inhabitants. These extremes are not the best forms of government, and certainly neither is evident in the state of Gaia. Mill's principle of liberty deals with a government preventing harms (Wolff, ch. 5). That's not the case with corporate interests in forests. Mill does not believe the no harm principle applies to those he calls barbarians, so he'd think nothing, it seems, of abrogating the freedoms and rights of nonhuman organisms or Indigenous people. Mill writes at the height of the British Empire and Victorian colonization, so his mindset should be different from a contemporary, ecological thinker, but that's no apology for someone purporting to stand against the status quo. Forests reached a level of maturity over eons, so they did not require external interference or governance. That peak has been waning due to human interference. The self-regulating utility of forests is important to planetary health and should be left alone with some regenerative help, as necessary, because of human-caused climate change. Mill suggests that some interests should be treated as rights and that we should act without harming another's welfare.

According to Darwin, evolution by means of descent with modification occurs roughly along variation, competition, and inheritance. Competition here is to be interpreted broadly and not as physical combat. Over time, as species vie for similar resources or territory, one species might die off or vary for survival and then pass along to the next generations those adaptations which enable flourishing. Competition in the forest works to the advantage of many species by selecting out and segregating ecological niches in an ecosystem. Industrial humans, on the other hand, do not operate this way and physically attack others or lands to exploit resources for financial gain with little regard for the effective function to climate health induced by forest organisms. Humans encroach on the interests and rights of nonhuman creatures to service their own self-interests, which tend to be non-sustainable, like mining, cattle ranching and farming to feed livestock, logging, etc. Gaia does not compete against itself but against humans. The rights of forest dwellers should be obvious. Equally, it should be clear how those rights are severely limited or abolished by human needs and desires. For instance, take a look at maps of the Amazon rainforest over the past 50 years to see how the diminished areas are immense because of human deforestation. Those abrogated rights include life, access to food and water, and free movement. Wolff (2006, p. 115) points out how these, at a minimum, are considered "human rights," but they apply to nonhuman organisms who have interests in their own survival and reproduction. Jeremy Bentham says there are no natural rights, and Mill follows that line of thought. Rights are not only derived from laws. There are natural laws, as shown by Darwin and his followers.

Rights for forest inhabitants are not abstract but part of nature's utility favoring the growth of healthy forests that sustain climate stability.

Mill says a right is that useful quality a society should defend (Wolff 2006, p. 117). In this way, it makes no sense to punish the health and well-being of innocent forest flora, fauna, and fungi since by doing so we reduce the ability of the world to thrive in a sustainable environment—all life support systems suffer. Following Mill and anticipating Martha Nussbaum who draws from Aristotle, individuals including nonhuman animals should be allowed to prosper. There could be assistance with laws to help maximize the success otherwise lost or at risk. The best approach for all would be to leave forests alone and intact. If some human encroachment occurs or natural disaster depletes old-growth trees, which can never truly be replaced, the best course of action would be for swift remediation of loss and reclamation of lands. With Mill (ch. 3), non-interference is crucial to freedom. This dictum applies, at present, only to humans and not to forest creatures or plants. In comparing the human and forest worlds, think how liberty is beneficial for the individual, society, and the state. While some state controls are necessary in a human nation, there is no such necessity to circumscribe the free range of forest animals. Mill grapples with the question of whether or not liberty always leads to improvement (Wolff 2006, p. 122), and again he's thinking only of a human and not the plant and animal world.

Human freedom mirrors that of animals in, for instance, play, learning, self-control, positive social behaviors, intelligence, etc. Forced control is not conducive to flourishing. While *mores* and traditions are important, the ability to develop is equally important. Among humans, there is no "innocent intent" (Wolff 2006, p. 126) in mining or logging forests for resources to fuel human desires and luxuries. On the contrary, consider the intentions and agency of forest animals whose actions keep the flora blossoming for their own and the planet's health. Human liberties often lead to forest harms; forest freedoms, on the contrary, sustain the earth's biosphere.

Karl Marx (1818–1883) and Friedrich Engels (1820–1895), Political and Philosophical Writings

According to cultural theorist Donald Hall (2001, p. 74), capitalism creates competition accentuating "self-motivated acquisition" where the proletariat exploited by the bourgeoisie now includes, in my line of thought, forest life forms. The labor of workers produces consumable goods that exceed the value of their wages but the capitalist profits (McInnes 1967). In this definition, economics is at the root of our climate crisis, too. We need to consider more deeply the capitalist ideology that shapes economic and material conditions. This ideology is perpetuated through popular culture like film, sports, and music and so can be changed by cultural evolution. For Marxists, there must be attention to and engagement with the material and economic conditions of life. They look for economic inequities that reflect social status with institutions hostile and will ing to destroy equity. Shift perspective a little, and you can apply this Marxist thinking to how industrialized humans treat forests. As Peter Burdon (2015, p. 5) says, human law "mirrors the values ... of the dominant class," which often sees the environment as irrelevant.

Capitalist hegemony has brainwashed social consciousness into mostly accepting an economic norm that harms nature and animals, but with cultural evolution this negative, materialistic routine seems to be changing. Young people who are inheriting both the climate catastrophe created by capitalists and the cognitive dissonance of most people are now fighting back. Marx addresses the question of resource ownership, allocation, and use for development, but he too, like Smith and Mill, seems to ignore any type of ecological economics (Costanza, et al. 1997). In human societies and economies, distribution of resources and wealth is directly related to rights. Some people and nations have more money and freedom than others through economic systems of oppression and domineering structures contrary to the mostly mutual cooperation in nature. Contrary to Marx's predictions, capital grew and did not die. Furthermore, the proletariat did not revolt. Over time, wages increased and surplus value or working hours were reduced in Victorian England. Marx predicted the collapse of capitalism, for in his day how could an economist ignore the conditions of the working classes in factories while apologizing for the status quo. Human life became a value reduced to labor. The doom to capitalism, according to Marx, is in how capitalists, seeking their own profit, would undervalue workers' labor. This devaluation has happened to forests for hundreds of years.

With capitalism, the social order is altered not necessarily through philosophy and mind but in the production and exchange of goods. Just as Marx and Engels saw social upheaval and change, now many people of all ages, primarily the young, are fighting for a new order of environmental and social justice with ecological ethics. Capitalism and the natural world are in opposition. We see conflict between the industrial capitalists and environmental ethicists. But this is not simply reapportioning wealth. Much more is at stake, like the future of some people and nations. Whereas Marx and Engels bemoaned the super structure of private property over workers, today, the overriding structure of capitalism threatens the environment that every organism requires in order to survive. Kept biodiverse with monetary incentives, private property could be part of the solution. As Marx said, it's not enough to interpret the world but to change it. Value in commodities consists of the labor, direct or indirect, to produce them. In his time, most workers earned close to a subsistence wage, and the extra hours were added to the profit of the business, what Marx called "surplus value" (Heilbroner 1961, p.132). The employer gets more value out of the workday than does the worker because capitalists control production and set working hours and pay, outside of unions.

Industrialists compete for these laborers and ironically use costly machines to increase the working pool and then lower wages. In competition with other companies using cost-saving measures, which affect output, profits could shrink driving the business into bankruptcy. None of this happened, but readers can see the analogy between the working class of Marx's day, their plight, and the flora and fauna of forests, which are similarly exploited. The difference is that those nations and corporations capitalizing from the many eons of work by forest creatures do not pay back. There is no equity. We need to revise political economy and society through cultural evolution, similar to the dialectical materialism (conflicting social forces) of Marx and Engels, to escape notions of industrial progress and create a new cultural paradigm perpetuating a sustainable ethics in the forest state of Gaia. Some individuals and groups are already engaged in this battle, as are corporate elements and parts of the United Nations, but we require a more forceful political approach. Resources in the forest are embedded in a free market to meet the needs and desires of inhabitants there. Resources are not owned, powered, or allocated by a central agency. There's no mismatch of the competitive processes of production and use. As Marx and Engels saw, economic and political problems ensue from modes of production (Feuer 1959, p. viii). Those problems don't exist in a forest state since there's a plurality of self-determined individuals whose cooperative actions engender a net result of growth.

According to Marx and Engels in their *Communist Manifesto* (Feuer 1959), a capitalist owns production and wage laborers. This notion gets to the heart of my argument, for the workers of the forest, whether microorganism, worm, insect, beaver, or bird, in spite of their ecological engineering that benefits all life forms, are at times owned by corporations and other powerful groups. As Marx and Engels note, emerging from feudalism, the bourgeoisie developed "conditions of oppression" (Feuer 1959, p. 8). The forest for capitalists has become nothing more than a means of production and exchange for the nexus of "cash payment" (Feuer 1959, p. 9). As they say, writing in the middle of the nineteenth century, in search of markets, capitalists infiltrated every corner of the globe for profit, and that's where forests became the new cash nexus. These days, it's mining for lithium. Capitalists create demands and seduce people to follow their lead into purchasing more material goods than they really need or even want. Ironically, the new capitalist frontier is profit in green technology responding to the elder capitalists who destroyed our climate.

Massive production and distribution have been facilitated by advances in machinery, transportation, and communication. Vast commercial networks have been established. Unlike the shared nourishment of tree roots, the capitalist enterprise places wealth, property, and power in the hands of a few. Marx and Engels talk about the plight of the factory worker and how he or she becomes an "appendage" of a machine (Feuer 1959, p. 14). Contrast this situation to the organisms of the forest who are, in effect, appendages of that superorganism. Forest flora, fauna, and fungi mostly thrive and do not suffer the alienation of some human workers. In the idiom of Marx and Engels, many human workers become mere instruments only to favor capitalists. On the other hand, forest "workers" from bees and lizards to beavers and deer are instruments of the self-regulating and sustainable state of Gaia. Colonizing capitalists have dragged forests into trouble. As Marx and Engels talk about the proletariat fighting against extinction at the designs of profiteering industrialists (Feuer 1959, p. 17), so too the forests of Gaia.

A difference is that unlike the lowest class of people in a highly stratified society, the forest inhabitants are not without property. They have their own lands and some sense of security and prosperity, which makes it all more shocking that it can be torn away by a platoon of bulldozers. Marx and Engels advocate the elimination of private property (Feuer 1959, p. 21). That's not my argument. Rather, let's call for the recognition of forest woodlands, with contingent wetlands and grasslands, as private to their rightful owners: fungi, animals, plants, insects, microbes, etc., who have worked those tracts of earth tirelessly for many millennia. Nevertheless, there are striking parallels to my thinking and what Marx and Engels say. For example, they point out that in capitalism, money supersedes individuality (Feuer 1959, p. 22). With the introduction of machines into the garden, plants and animals lose their sense of presence, self, and purpose.

Marx and Engels call for countries to unite in action on behalf of the working class, as we now need for the flora and fauna of Gaia. A central idea in the *Manifesto* is that human consciousness changes with the material effects of existence (Feuer 1959, p. 26). In his *Critique of Political Economy*, Marx says that human life is not determined by inner consciousness; individual and community minds can be directed by material and social existence (Feuer 1959, p. 43). Learning from animals we know our needs can be quite simple, evidenced in people who downsize, live in tiny houses, or have reduced their desires and waste. Capitalists can easily dictate consumer needs, which affect manufacturing and profits. As Engels says in "Socialism: Utopian and Scientific," capitalism by definition enlarges and concentrates production for profit to benefit a few (Feuer 1959, p. 92). For Marx, the state-oriented society with a feudal mentality determines and cripples one's consciousness and existence since there's a loss of self under the burden of a controlling power (Feuer 1959, pp. 360, 362). If we can alter human consciousness in favor of forest sovereignty by emphasizing less desire for resources stripped from those lands, then a cultural shift could occur that favors cessation of deforestation for everyone's welfare.

In his main work from 1867, *Capital* (ch. 10), Marx discusses how capital forces every ounce of production out of workers each day. This is what happens to forests, which are squeezed dry of resources by nations and corporations. Conversely, in their own cycles, and unless disrupted by natural or human disasters, forests, wetlands, and grasslands are self-sustaining and flourish. Marx uses the metaphor of ravenous hunger to describe the capitalist's appetite for labor and hence profits. That method of production is not sustainable when it comes to abusing nature's economy. Marx cries out about the degradation of workers in Europe and slaves in America, how their lives contra Kantian ethics are merely means for the ends of capitalists. We see this too with the flora and fauna of forests. There are machines that can cut and strip a huge tree in minutes. What of the microorganisms, insects, and birds who call that tree home and who might be ground up in corporate gears.

Marx goes on to say how capitalism devises propaganda and "reasons" to lie about the condition of workers (ch. 10). Today, some nations, politicians, and corporations will try to make their work of drilling, logging, and mining appear innocuous. We are losing alarming tracts of Amazonian rainforest in Brazil, up to 170 hectares for beef cattle and food production for them and other livestock (Attenborough 2020, p. 97). The whole process of machine deforestation, feeding herds with grains that could have been used to nourish people, accumulating offal, and wasting water is staggering. Marx states that the maneuverings of capital are in a never-ending process of taking away property and means of production from workers (ch. 26). So too with forests, for with every hectare deforested, there is loss of biodiversity that captures carbon, refreshes air, and collects water. Divorcing flora, fauna and fungi from servicing their inherited ecosystem niches is unsustainable for the planet. Just as capitalists can expropriate workers, they can also demean and diminish forest workers and seize their subsistence. Individuality is lost to the oily contraptions of capitalism since the worker's labor becomes another's profit, Marx might say (ch. 32). A consensus of scientists (e.g., Jamieson 2018; Andrews & Monsó 2021) agree that animals are individuals with distinct personalities, so to deprive them of habitats through which their actions mitigate climate chaos is unsustainable and unethical. Essentially, corporations appropriate through proxy nation-states and steal the property and individual means of production of forest inhabitants like capitalists during the industrial revolution. Forest flora, fauna, and fungi work in social cooperation for the greater good, which is antithetical to capitalism's emphasis on profit and the accumulation of wealth for a few as proclaimed by Milton Friedman (1970).

From a Marxist perspective concerning liberty, there might be a difference between the state and civil society. For instance, a state could emancipate its people from certain political restrictions and view them as equals (Wolff 2006, pp. 128–29). At the same time, there's an underlayer of bias and hatred that erodes the state. We don't see this perversion in the forest, but among humans it leads to and justifies discriminatory laws where those in power benefit over others, as seen in forest destruction. A free and liberal state with a focus on capitalism and property ownership enacts policies detrimental to forests and consequently the global climate. The assumed rights of the state separate themselves from forest dweller rights, degrading climate health. Marx was not a proponent of equality because he viewed it as a flawed ideal. Marxists are less about distribution, which can always be unjust, and more about transferring the means, control, and ownership of production. From one angle, a Marxist claim would transfer climate health production to forests and eliminate the environmental virus incurred by humans.

For Marx, society would not involve economic conflicts because there would be no scarcity. It's not about equalizing property as per John Rawls (1999) or distributing social goods based on group values as per Walzer (1983) but eliminating private property. Because of evolved adaptations and ecological niches, no one rules, and none are taking advantage of others in the forest, so it's close to social ownership. Some might say that predation is a problem; but no, since that is a means of sharing resources and controlling populations without depleting them and why forests need to be expanded. In the Gaian system, predation is a metabolic exchange. No one is exploiting surplus value for profit from another's production. Rather, labor and profit are shared. Products are not created *ex nihilo* but sustainably from the feeding practices and excretions of others, whether bacteria, insects, fungi, flora, or mega fauna. No single political theory dominates in the forest state, and we can see that in line with Marxism there is no alienation of its citizen workers. Technically, with equal access to resources, no one is exploited unless by humans.

Borrowing ideas from political philosophy, as with Marxism, forest animals take what they need from abundance according to their evolved idiosyncrasies. There's an equal use of all resources. Likewise, in Marxist society resources are distributed so that people can achieve self-realization. This means there is cooperation against alienation in libertarianism or liberalism where resources are hoarded and depleted. Apparently, Marx thought only humans could produce collectively for their higher benefit and life value (Kymlicka 2002, p. 193). In the forest, resources are shared, there's community life, and all organisms contribute to many lifecycles. In Marxist thought, workers are crucial to generate social change. Forest workers as environmental engineers, assuming there are more of them on a planet with extensive and contiguous forest cover, can help societies who would otherwise suffer damage from climate change.

Continuing with the analysis of political philosophy, let's now turn to the more recent and at times opposing voices of liberal John Rawls, libertarian Robert Nozick, and pluralist Michael Walzer in the next chapter expanding outward with related considerations in Chapter 4.

Chapter 3

Interpretations of Political Theory and Implications for Forests

In this chapter, I'll examine commentaries on the political philosophers outlined in the previous chapter in comparison to influential contemporary thinkers like John Rawls and Robert Nozick with a note about Michael Walzer. Since political philosophers focus on shared governance of and by humans, animals are excluded, say Sue Donaldson and Will Kymlicka (2016a, 2016b). Animals are deemed cognitively insufficient for political participation, and Donaldson and Kymlicka find this anthropocentric thinking inadequate. At its core, political philosophy deals with conflicts of coercion, power, and legitimacy. Over the course of millennia forest areas, on the contrary, have proved a high level of self-sufficiency. Even so, nation-states duly authorized to wield power engage in attacks on humans and animals alike, to say nothing of deforestation. Animals, whether domestic, farm, or wild, along with forest flora, wetlands, and grasslands are everywhere, but we don't treat them with the justice they deserve. Again, in light of the proposal in the Preface, we need community members (especially Indigenous worldviews), biologists, lawyers, and philosophers who can speak for wildlife since plants and animals are not truly represented in the Western political tradition.

John Rawls (1921–2002), *A Theory of Justice* (1999)

Rawls, whose book was originally published in 1971 and then revised, follows Kant's rational individual rising above self-interest (Kuehn 2001, p. 33). There are protections of freedom, equal opportunity, and benefits to the disadvantaged. Rawls and the Enlightenment tradition emphasize a freedom and equality of individuals as rational agents, but this has been challenged by communitarians and feminists who find the self in a social sphere (Christman 2002, p. 5). Rawls moves away from a universal moral theory of justice for the singular person, in *Political Liberalism* (1993), and acknowledges acceptance of conflicting and combative political ideologies. Forests appear to allow a liberal diversity of behaviors, even if some actions, as in feeding, seem by human standards restrictive. There's an even-handedness in nature overall that we do not see in autocratic and even some democratic human societies, for example infringement of rights for marginalized people. No organism is excluded from the rights, property, and liberty in nature's bio-network but has evolved and adapted an ecological niche.

For Rawls, justice as fairness or the equal distribution of rights and opportunity, especially for the disadvantaged, should stem from basic political, social, and economic institutional structures (Farrelly 2004). Compare this proposal with a forest where no one is really disadvantaged, as shown in Chapter 1 herein. Rawls is more concerned with right than maximizing utility. We can see how the mechanisms of capitalism favor the well-being of some people in wealthy nations and what they perceive as good over fairness for others and forest life. As Rawls might suggest in this argument, the fate of forest flora and fauna has been treated as an administrative decision by a few people for the benefit of others with no just regard for plants and animals. Rawls would promote a duty of mutual aid, as we see in the forest, that positively affects many life forms.

Drawing from Locke, Rousseau, and Mill, justice for Rawls is fairness where free and equal members of a society honestly cooperate with and contribute to an economic system beneficial to all (Wenar 2021). Any set of human rules imposed upon forests has not been endorsed by the flora, fauna, and fungi and so is not legitimate. Along with Locke and Mill, Rawls envisions people as capable of tolerance. According to Rawls, what Nozick lays out in *Anarchy, State, and Utopia* (1974) is not liberal justice since libertarianism excludes the freedom of many by excessively benefiting the power of a few. John Christman (2002, p. 6) notes how the word liberalism means human rights in constitutional democracies and "popular sovereignty." Forests represent multicultures reflected in modern nation-states, but there are political forces that prefer a ruinous monoculture, which would be the death of human society and nature. Just as there should not be rights for only one group (e.g., white men) at the exclusion of others, we cannot have rights that favor only one species or a fantasized version of nature (e.g., zoos). As Christman suggests, there are immediate and immense possibilities in politics we could embrace, mainly if the benefits are widespread, as in expanding and preserving forests.

Relying on publicly accepted values and standards, Rawls wants to know what most members of society would reasonably believe about liberty and equality. Hinting that equal justice is reserved for rationally thinking moral persons, he discusses how there is no "natural feature" that equalizes all humans (1999, p. 444); but nature favors equality, he admits. I'm trying to apply that line of ethical thinking to forest economy. In Rawls' conception of justice as fairness, there's no privilege but rather an equity of advantage in distributing social goods. Walzer (1983) follows a similar plan but envisions distributive justice in overlapping spheres, with the sphere of membership as pivotal: who's included and who's excluded. For Rawls, there's mostly one principle of justice that embraces equal liberties and opportunity where great benefits devolve to the least advantaged. Ideally, some of Rawls' ideas reformulated and applied to forest protection could be codified into state constitutions, as has happened in the U.S. (e.g., Article 14 of the NY state constitution). Informed citizens can make responsible decisions about available resources and are not always enslaved in thought or action by a powerful few.

Moral and political theories must answer two fundamental questions. First, what are the demands of morality? Second, why are we obligated to obey them? The demands of morality or even politics are fixed by agreements that humans make to regulate their interactions. We should obey these agreements because we have consented to them. A social contract provides an answer. The state of nature is nonpolitical, whereas the civil state is political. The social contract is a voluntary agreement of all people. Some main assumptions include the following. Political relations lack a natural basis and are artificial by consensus. In the state of nature all people are free and equal, and there's no higher authority with power to command or responsibility to protect. In these assumptions we agree to surrender some of our natural power to the state. Civil power is held in trust, and obedience is conditioned on the state effectively fulfilling its function of protection.

Justice as fairness is not a negotiable quantity where it might be with a utilitarian thinker. For Rawls (1999, p. xi), "justice as fairness" is an alternative to utilitarianism that only approaches an ideal as a frame for moral sensibility (1999, pp. 43, 46). Commitments should be tied to virtues. Utilitarians care about the general conception of the good, the opposite of justice since the focus of right is the sum total of welfare. Sacrificing one's rights, as a utilitarian might suggest, for the sake of the rights of many is not just. A just society must secure rights for all. Social cooperation is an advantage for everyone with no "arbitrary distinctions" about rights, a balance (Rawls 1999, p. 5). Social cooperation is ongoing, not static, and concerns rights and how advantages are dutifully distributed. We can create a concept of justice through advantage and disadvantage. We can create a contract in a way that distributes advantages fairly. There might be unfair assumptions in that people are rarely impartial regarding their own positions. All of this applies to forests in several ways. First, most people will agree that undue force used for unfair gain is wrong, as we see in how nation-states employ corporations to destroy forests. Second, most people would be willing to correct an unfair advantage that profits a few while harming the health and well-being of many.

For Aristotle, justice is in terms of actions from motives and includes the virtue of restraint in taking advantage of someone or property or by denying what is lawfully another's (1952, 1129b–1130b5). Unlawful seizure can even be sanctioned by elected officials. For Rawls, justice is made or agreed upon, not found by a few people in power. This is not the case when humans take over forests. Following Aristotle, human institutions have gained material and social advantages from the forest state unjustly. Rawls says that when he and others talk about the state of nature it's hypothetical. Rather, nature is alive, real, feels pain, and reacts. In a social contract where the parties agree on principles of rights, distributions, etc., justice might demand that elements lie behind a veil of ignorance, so people don't know of another's income, status, or intellect, says Rawls (1999, p. 11). Yet there is no such instance of being "similarly situated," despite Rawls, between humans and nature where, ironically, humans believe themselves exceptional and entitled to take whatever they want.

Rawls says there should be an equality of rights and duties, and any inequalities are just if there is even compensation, especially for those disadvantaged (1999, p. 13). Some need not suffer social or economic hardship so that others prosper. There could be cooperative division of advantages. Some are better off by natural endowments or fortune. According to his difference principle, there should be fair equality of opportunities where the least prosperous benefit. Rawls' "difference principle" means that inequality is justifiable if it helps those disadvantaged (1999, p. 68). Apply this principle to forests that contribute significantly to all organisms and yet are undervalued but for the profitable resources and systems that give rise to our planet's health. A question of interests and distributions is who and how many equally benefit considering responsibilities and authority. Some inequalities might be just, but that's not the case with how we have treated nature unfairly without any virtue of restraint, as Aristotle would declare. Where one might say forests are, in human terms, least advantaged, others might claim that humans are least advantaged without the necessity of ravaging forests.

Rawls wants to put aside moral points of view that are arbitrary. He admits (1999, p. 15) that his justice as fairness theory does not currently include human relationships with animals or nature. He admits that "animals" do not have the same status as humans in terms of rights and equality, in spite of his nods to evolutionary thinking (1999, p. 442), but should be treated humanely. Animal welfare and humane killing are strange preoccupations that have assisted the promotion of social and business ills under discussion. There has been impartial judgment and unfair advantage when it comes to establishing relationships with forests. Gaia has not been represented or consulted except by proxies in conservation who are typically out-gunned by well-financed lobbyists and commanding politicians. There is little negotiation or, as Rawls says, "equilibrium" (1999, p. 18) in many one-sided deals. However, as he suggests, the principle of justice is revisable, as recommended here by establishing the sovereignty of Gaia with due representation and restitution for it to thrive.

That which is right produces the best for the greatest number in society, according to utilitarianism. The problem with this line of thinking is twofold. For one, the society of nature has not been considered or consulted about what is "right." For another, since the twentieth century deforestation and other forms of destruction have not benefited society, evident in our climate crisis. Adam Smith veers toward utilitarian ideas when he says a rational, impartial spectator sympathizes with others as if with himself. The problem with such thinking, for Rawls, is that individuals are lost in a system that seeks to maximize satisfaction for many (1999, p. 24). If all individuals are lumped into a mass that supposedly provides justice for many, each person is not truly afforded a right to justice. Rawls sees utilitarianism as teleological and in opposition to contract theory, which is deontological. Borrowing his words, justice as fairness does not separate "good" and "right" and does not see right as something to maximize good (1999, p. 26). Following utilitarian logic, the greatest pleasures for most people have come from eating cows ("beef") and burning products made from forest resources, but those human desires are not fair or just to the flora and fauna of Gaia. We can culturally evolve to terminate values that are socially destructive, and that includes deforestation and the consequential climate ills from that practice. So, for Rawls, in this logic, right precedes good.

A theory of justice applies to public systems, legal institutions, and capital markets, not the same for individuals in certain circumstances. Individuals will question the morality and social ends of rigid rules. Here are Rawls' justice principles: 1. equal rights of access for all to "basic liberties" and 2. inequalities could be reasonable only when all are advantaged and in "positions ... open to all" (1999, p. 53). Later in the book, he expands on inequalities to say the least advantaged should receive the greatest benefits; equality of opportunity should be available to minimize social and economic inequalities (1999, p. 266). Of course, all of this is explained in human terms. What we consider socially valuable, like freedom, opportunity, and dignity, could be equally distributed, but isn't, either in society or as applied to nature. Taken as a whole, we might adjust fairness as justice to acknowledge the forest state of Gaia as integral to our "social" design. By these definitions there is no justice for forest creatures since the destruction of their habitats and the distribution of what remains, land and resources, is not to their benefit. An alternate distribution scheme Rawls alludes to is easy in terms of the forest. That is, expand and then leave alone vast tracts of woodland, wetlands, and grasslands to their own climate healthy functions.

Rawls makes reference to Pareto optimality in economics. With any redistribution some party will benefit and another will lose (1999, pp. 58–59). Rawls' concern is that any focus on efficiency is not justice, reminiscent of Aldo Leopold (1949) minimizing expediency of land for human use. Regarding modern humans in forests the Pareto principle is true, whether one sees it as normative or scientific. However, the point is that left to their own devices, forests are quite efficient and sustainable for all without consideration of optimal expediency for a select few. Following his line of thinking, human actions upon nature do not permit the liberal flourishing of talents evident among forest flora and fauna. Human market and social contingencies deplete, one might say in the lingo of political economy, the natural distributions that would sustainably occur in nature.

At this point in the discussion, it might not be difficult to see how three ideas from Rawls' political theory (1999, p. 92) converge with the argument about forest sovereignty.

1. Society is a "union of social unions," which is true in the forest. Rawls says that while there are conflicts, society represents an "identity of interests" (1999, p. 109), also true of forest wildlife and vegetation.
2. A class system can break open with diversity in a free society. One's own labor ought to result in fair gains. Rawls notes (1999, p. 151) people have "final interests," which can be said of forest organisms, too.
3. All abilities can be used for greater social advantage; true, as noted with Lovelock's (1979) Gaia hypothesis. Competitive markets to date have trumped cooperation, and while they have diversified, nature has suffered. Utilitarians stress benevolence and sympathy in their system because in counting sums there are losers, along with little respect for others, and not "reciprocal advantage," says Rawls (1999, p. 154).

Rawls fundamentally defines liberty (1999, p. 179) as "equal citizenship" for all to achieve their ends without harm to anyone else. In human terms, some are far more advantaged than others, for example by wealth or status at birth. Rawls says that adults should make rational choices with future generations in mind, as if those descendants would decide rationally their preferences, especially for liberty. Even for Mill, liberty is not equal for all since utility is basically teleological. We should only agree to principles whose outcomes we'd accept, but that's difficult if we are not rationally impartial (Scanlon 1982). It's doubtful a hardened capitalist can put aside his self-interests and become sympathetic to forest and social needs. Fairness as justice tries to settle conflicts with social equality in mind, but this does not always occur through utilitarian impartiality (Rawls 1999, pp. 162, 165). In the common interest, and in line with toleration, a government must maintain order and security. This idea contradicts what's occurring in forests. In order to maintain public order, Rawls' phrase, Locke and Rousseau are willing to put some restraints on freedom. Liberties imply responsibility, so there could be control on how capitalists, enabled by politicians in some nations, defy the liberty of forest wildlife.

Looking to Kant, moral sense as a contract doctrine is free from circumstance or contingencies (Kuehn 2001, p. 281). Moreover, via Kant (*Metaphysics of Morals*) Rawls suggests one's character should be determined by acts of social morality and not simply by fulfilling selfish pleasures (1999, p. 224). Following Rawls, why not guide political economy based in justice toward the public good. That realm of justice would include forests, wetlands, and grasslands. Institutions cannot be determined only by current wants but by anticipating future needs. Economic systems can shape current consumer trends that anticipate future directions, so there is a moral component to be considered. There can be moral problems in political economy, for example taxation and social spending, production and distribution. Only in passing does Rawls mention how free markets pollute without regard for any social good (1999, p. 237). Common sense dictates there can be justice between generations. It's not fair for one generation knowingly and deliberately to maximize for its own benefit all resources leaving little for the next generation. That does not happen in Gaia's nature. Worse, there's the dilemma of harming the environment and expecting the next generation to replenish natural resources, if even possible. Institutional establishment must be arranged and managed to deal with changing conditions in the present and for the future, Rawls would concur (1999, p. 254).

Why can't politicians agree to prioritize care over privilege or status, health over cash wealth, resource sharing and conservation over property or political power. Negative attitudes, Rawls points out (1999, p. 264), from the burgeoning industrial nations in Europe of the long eighteenth century set the pace for the crisis we now face. Rawls' principles consistently favor the least advantaged, so that should include, in a revised attitude that is not financially anthropocentric, the flora, fauna, and fungi of forests. Harming the forests is disadvantageous to Indigenous people, to poor people in developing nations, to people in climate sensitive areas, and to all organisms present and future. In sum, the state of Gaia is the territory that affects all nations if harmed. Therefore, it only seems ethical to grant forest sovereignty and endorse its immunity from human incursion except for existing Indigenous populations since once left alone it is naturally sustainable and a critical factor in planetary health.

Again, with reference to Kant's *Metaphysics of Morals*, Rawls talks about mutual aid and the benefits of helping others by spreading support (1999, p. 298). For humans, beyond actual assistance, mutual cooperation is a signal of trust. Publicly recognized standards are different than those of the Victorian era or after World War II. National economic growth for the common good might have been acceptable then, but there were few promises made about protecting the environment. When environmental laws were eventually written, lawyers helped large corporations avoid liability. This is still true, though some businesses bowing to social pressure are engaged in public relations and green programs, some of which are deceptively greenwashing. Nevertheless, given Rawls' principles of justice as fairness, political and corporate actions are not totally in line with forest preservation. Because of the actual climate crisis, there are new obligations on representatives in government who must act in good conscience and not for financial or political gain. Voters need to understand the gravity of environmental conservation laws and how their representatives will vote. Rawls goes on to say that the majority is not always right (1999, p. 313), whether among the populace or elected officials, as is obvious concerning the environment.

What is good relates to the difference principle that favors those least advantaged. A rational plan for the future should include not only action items but also what not to do. Rawls cites an Aristotelian Principle as follows. Whatever we excel at gives us greater pleasure the more complex the tasks become and we succeed (1999, p. 374). Consider, for example, athletes and poets. He goes on to say how this Aristotelian Principle is a means of motivation. This is clear from how more wide-ranging in complexity deforestation has become. The principle goes against E. O. Wilson's non-interference policy. As an institution, corporation, or nation matures in abilities, simple tasks are no longer satisfying. Perhaps too optimistically or naively, Rawls sees, eventually, an equilibrium; but without constraints there's no end to loss of forests, wetlands, or grasslands. Going against what Rawls says elsewhere, given the deepening scope of deforestation and climate crisis, this Aristotelian Principle is not rational. Ecosystems find balance, not seen in human needs and desires. Where flora, fauna, and fungi mostly adapt sustainably, humans mostly choose poorly. The naturally wiser consists of wildlife, not most people in industrialized nations. Natural selection did not necessarily favor organisms with this capacity of ideal self-realization, as even Rawls suggests (1999, p. 378) since many life forms evolved simple and clean ecological niches for livelihood. If anything, the Aristotelian Principle is anthropocentric.

Justice principles don't work if they cannot motivate people to create and maintain impartiality and fairness. Rawls' book attempts to show the psychological realism of justice as fairness in a system of equilibrium across a well-ordered society. He occasionally makes nods inadvertently to biology as follows. Human good nature is innate (1999, p. 403); family structure is basic (1999, p. 405), though this can be heavily qualified in the nonhuman world; parental care (1999, p. 406); childhood influence and learning (1999, p. 407), though this too might require qualification. In Chapter 8, Rawls discusses universally stable and natural moral sentiments like justice and fairness in relation to individual moral emotions and feelings (e.g., guilt and shame). What must be held accountable are the moral sentiments, attitudes, and intentions of corporations and political entities. The moral feelings of individuals in the current climate crisis can't be ignored, either. One large problem area is how institutions contribute to moral psychology through crafty advertising, propaganda, or dodging negative news. There's a massive oil spill, but we cleaned it up, and you'll have gas for your car. We kill Indonesian biodiversity when we deforest for palm oil or sugarcane plantations, but the farms provide local people with work and you get cheap oil for your potato chips, food products, shampoos, etc. Even Rawls questions the moral attitudes and feelings which helped us evolve (1999, p. 433). The more important consideration is how we can culturally advance away from material consumerism and heighten our sensitivity to forest preservation and overall wildlife conservation. Rawls cites Mill's notion of the perfecting mind harmonizing with others (1999, p. 439). This is a far-fetched idea in a world of 8 billion people, many in competition and conflict. However, this means Mill is veering away from utility and looking at the difference principle, which we can apply to forest organisms.

Rawls' principle of justice as fairness along with the veil of ignorance, if accepted, means that people act morally as individuals without influence from the opinions or beliefs of others (1999, p. 452). What I'm suggesting is that the epidemic "virtues" of capitalism with their Enlightenment traditions could be replaced by a focus on preserving Gaia. Let's put limits on private interests in public goods. For example, consider the invented benefit of yet another massive highway or railway that slices in half a forest or prairie with no provision for wildlife crossings. If public goods are benefits for all and from which there's no exclusion (Miller 2003), then wildlife should be included. Considerations of "social union" (Rawls 1999, p. 459) are critically important in answering questions, but social union is always anthropocentric. Yet we see social integration in nature where individuals and different species interact and cooperate to create a net effect of sustainability in the forest. As with eco-engineering this social combination does not have to be conscious or intended. Just look at the number of benefits to see how it is effective.

A problem is that people with inflated notions of "humanity" want something from the natural world that is far different from the cooperative ethos we see in forest flora, fauna, and fungi. There's very little social ethics regarding human action in the forest. Would we want invaders destroying our homes and infrastructure? Contrary to Kantian principles (Kuehn 2001, p. 280), benefit for all is not upheld. Liberty and economic advantage are at odds, he might suggest (Rawls 1999, p. 475). With some irony, forest organisms make decisions about what is ecologically good in near isolation. Exceptions to isolated behavior include the mother-infant dyad, if there is one for the species, and group dynamics and division of labor, if that applies. Though many species are in sympatry, working environmentally the same portion of earth, they are probably alone. For humans, lack of independence means physical and psychological control by those in political and corporate power. Rawls spends a lot of time in his final chapter saying how cooperation with others in justice is part of a public good, repeating the phrase "social union" (1999, p. 500). This is not utilitarianism, which ultimately sacrifices the welfare of some for the benefit of many others, what he calls a *hazard* to freedom with a question of priorities. For the public good, then, forests, including adjacent or interior wetlands and grasslands, could be granted sovereign immunity from any human intrusion.

Robert Nozick (1938–2002), *Anarchy, State, and Utopia* (1974)

Robert Nozick is even more anti-consequentialist than Rawls, promoting a minimalist state except for ensuring individual property rights. According to Nozick, there should be no state power to redistribute income or to enforce moral norms. Per Locke in his *Second Treatise*, section 6, there are natural rights that precede any state, and so no one can be asked or forced to act in a way against these rights even if it's for a social good. Individuals have rights that precede any claims made on them by political or moral norms of a state. In a natural state one is free but should not interfere with, restrict, or transgress another's freedoms, unless willfully waived. Like Rawls, Nozick is a critic of utilitarianism, which roughly sees persons similarly in terms of minimizing/maximizing costs/benefits. Rawls is more contractarian; for Nozick, natural rights precede any social contract. Rawls and Nozick both opposed utilitarianism since advantages must occur within a life and not necessarily across society (Mack 2022). Individuals are not to be used as a means for or by others without affirmation. In Chapter 7 of *Anarchy, State, and Utopia*, Nozick insists that one cannot take the outcomes of another's labor without consent and compensation. Nozick argues for a minimal state that encourages peace and community protecting the freedom and rights of individuals.

Apparent by now, virtually all of the preceding paragraph should apply to the plants and creatures of the forest who have natural rights to those lands, whose property should not have been confiscated or destroyed, and who should along with Indigenous inhabitants be left alone to continue cultivating those large ecosystems for planetary health. As Thomas Nagel (2013) puts it, Nozick espouses a minimalist state with liberty for individuals without state-imposed social norms or welfare. While Nozick leans heavily toward Locke's notion of natural freedom, he does not see, like Locke, a large state arising out of consent from individual rights. Moral norms are outside of institutions, a major break from Rawls. Of course, it's well-know, later in *The Examined Life* (1989), that Nozick tempered some of his positions, but for the purposes of this discussion I am accepting his argument in *Anarchy, State, and Utopia* since it is such a highly regarded, influential book used for years in college courses and translated into many languages. Nevertheless, the failing I see is that as with Rawls and others, Nozick is anthropocentric and assumes, even in the "state of nature," that only humans are worthy of consideration.

With Nozick there is no Rawlsian distribution favoring the least advantaged. There are no spheres of distributive justice in a pluralistic state as per Walzer. A minimal state as imagined is focused on security of persons, property, and contracts. Pursuant to Walzer, there is no membership for security; safety and welfare are distributed evenly without preferential dominance. For liberal thinkers like Rawls, freedom and equality are as one; for libertarians like Nozick, it's only liberty that matters (Farrelly 2004, p. 34). Freedom and equality are not exactly the same; animals in the wild are free but hunted, and pets are like family members but not free. According to Nozick, if a state coercively enforces distributions, that is unjust; however, we find noncompulsory and equal resource dissemination in nature for the advantage of many. Nozick favors absolute property ownership; history shows how that has failed many people across nations. Attempting to correct Hobbes, Nozick says that in movement from nature to the minimal state there would arise protective agencies (Farrelly 2004, p. 37), but those could be confrontational with no guarantee rights won't be violated.

One can understand how Indigenous worldviews might align with Rawls and liberalism, but what about libertarianism. Indigenous people, contrary to libertarians, don't see land as private property but as communal ownership; libertarians advocate for free-market resource allocation, whereas Indigenous people share resources among themselves and with wildlife; the cultural practices of Indigenous people are self-determining as a group, but libertarians would prefer individual autonomy with minimal governance. We can see here that wild flora and fauna are more in line with Indigenous thinking; even among plants and animals, genes respond to the total environment of animate beings, climate, and inanimate objects, so there's no total self-determination.

For humans, there are questions as to whether protection is for liberty and property (Nozick) or for equality and welfare (Rawls). Nozick erroneously believes humans alone are capable of and naturally entitled to self-ownership as a moral right. A model of the minimal state Nozick prizes where respect for individuals is part of the evolutionary process appears in nature. At the same time, the natural world abounds in mutual aid and welfare for others in just proportions as needed, even if it's unconscious. Contrary to Nozick, absolute property rights do not exist in the real world. In the wild, large animals stake territories, but others cross those lines and use that space. Large trees are eco-habitats. Smaller creatures and organisms, including ground-level plants, have evolved circumscribed ecological niches; but here, too, these places can be used by others from parasites to unrelated species as seen in how tree roots and fungi work together.

Nozick (1974) begins by asking what the state is permitted to do concerning strong individual rights. There should be limits to a state's coercive force, especially as it might check freedoms. Nozick, in this way, says the state is "immoral" since it "violates" individual rights (1974, p. xxi). Nozick prefers Locke's state of nature where, he says, rights are not violated. In another view, the state of nature is not anarchy and is preferred to the capitalist-industrial state evidently intent upon destroying nature. Nozick refers to a state as an "apparatus" that will, from the perspective of an anarchist, immorally limit one's freedom (1974, p. 6). He sees the state of nature as nonpolitical perhaps since Aristotle finds only humans as the consummate political animals. Frans de Waal might disagree based on his groundbreaking book whose title says it all, *Chimpanzee Politics* (1982).

A given among cultural anthropologists is that while there will always be cheaters, they typically don't prevail in a group. Wolff (2006) admits there are drawbacks to a state, for example control of many by a few. However, he suggests that large groups in nature are politically unsustainable and require a government over what he and other political philosophers would call a state of anarchy. Anarchists prize liberty and see governments as a threat to freedoms through coercion and interference (Schmitt 2009, p. 91). Anarchists, following Rawls, see unjust inequalities like power and riches for select groups. These kinds of differences don't exist in nature and have been created by humans as they plunder nature for mineral and material gains. Anarchists would be happy to learn that there is no real government in nature and all are free; yet, there is no chaos in the forest. Rather, forests are adaptively ordered and self-organizing systems. In Wolff's (2006) references to anarchist-scientist Peter Kropotkin and mutual aid, he seems to offer a simplistic evaluation of Darwinian struggle as physical competition. Wolff recognizes selfish behavior in cooperation and how ultimately cooperation produces a better outcome for everyone. Yet he does not acknowledge that cooperative behaviors are evolved adaptations in Darwin's world. Following Hobbes, Wolff says the seventeenth-century philosopher would note how aggression and competition could be justified as rational, spreading through a group. Again, Wolff's focus is only on human cultural behavior. For example, he says not all people throughout history have been "good" (2006, p. 31) since there's ample evidence to the contrary. This good/bad attitude is anthropocentric, but he says there can be "social control" (2006, p. 31) of behaviors without any formal state. We see that organizing factor among social animals from ants and bees, wolves and elephants to primates.

There are many social insects with evolved divisions of labor, as well as pack and herd creatures. So, it's a bit puzzling when these philosophers talk about a state of nature as if only specially created. Later, Nozick refers to "rational self-interest" to create citizens and not victims (1974, p. 17). He seems to defer to Adam Smith's invisible hand with individual gain over Locke's civil society by mutual consent. Nozick invokes F. A. Hayek several times in his book, a thinker who favored a free-market economy over socialism. Nozick, following Hayek, sees capitalism as service to others (1974, p. 218) where Rawls would see in free enterprise unequal distribution. In terms of moral constraints, forests, like animals on farms or in most zoos and all labs, have few if any rights. Compare those restrictive situations where nonhuman life is used for human purposes to animal sanctuaries where pigs, horses, goats, cows, sheep, dogs, cats, etc., roam freely. While some organisms can be forced into labor or entertainment (horses throughout history), most forest beings contribute to each other's welfare with kin altruism or interspecies mutual aid through evolution. There is no top-down autocracy or capitalist monopoly. Instead, humans use forest life as a means to their own ends, essentially pitting organisms against each other as habitats shrink and feeding resources disappear. Nozick (1974) talks much about individual liberty but affords none to nonhuman creatures since he seemingly views them as resources, though he does not condone eating them.

Humans fully intend to aggressively exploit forest animals, their habitats, and their resources in spite of obvious environmental damages. According to Nozick's political philosophy this violation should be considered wrong (1974, p. 32). Though apparently vegetarian and opposed to unwarranted aggression, he draws a Nick Zangwill (2021) claim as follows. It doesn't matter that we eat animals; they've benefited from us in the short life we've given them (1974, p. 38). Then in reverse he goes on to note, contra Zangwill's moral obligation to eat meat, that killing animals for food is an aggressive violation of their claim to life (1974, p. 38). Nozick is an important thinker, and Josh Milburn (2017) extrapolates from him an amiability of moral obligations and valued respect for animals, closer to his ethical and not necessarily political philosophy. In looking at things from a utilitarian perspective, animal agriculture is no longer a benefit for social good with all of its pollutions and causes leading to climate change. Nozick claims that a state equals protection for all, not a few (1974, p. 51), but this ignores the forest state of Gaia that protects organisms invaded by humans. The state also has power to punish, though we don't typically see that in, and not often enough on behalf of, Gaia. In *The Second Treatise*, section 95, Locke says some may choose to exercise their liberty in nature and not in a civil state. This sets up a conflict for Nozick because he thinks only in human terms. Meanwhile, the liberties of the forest negotiate space and resources through evolved adaptations and niches. In discussing prohibitions and risks Nozick hints at compensation. There are social and psychological costs related to climate change (Doherty & Clayton 2011; Hickman 2021) for which there's scant compensation.

To date, there's no just or fair restitution to the forests, though of course some people (Indigenous) and nations (developing) suffer more than the capitalist-industrialized states. There could be physical borders with severe consequences of compensation and retribution at this point of the climate crisis. Nozick claims as problematic how in a state of nature there's no central decision maker (1974, p. 74), but that's an advantage. He brings this up in terms of fear, risks, and compensation, but he's imposing faulty human structures on the natural world. He views everything transactionally as an exchange with a winner and loser, but that's not quite how the systems of biology evolved in nature. In his line of thinking, a free society would continue to permit deforestation without any fair or just recompense to Gaia (1974, p. 78). Yet, according to his theory, compensation is in order since there has been harm. Moreover, he'd argue that prohibiting actions harms the actor and inhibits free society (1974, p. 79), but that can be viewed differently whether one considers corporate or wildlife activity.

At one point, Nozick raises the issue of pollution, but discusses "only effects on property" (1974, p. 79). That would be human property in his opinion. He literally says all polluting activities should not be prohibited, especially if there are benefits that exceed costs. He raises the question of whether polluters should pay damaged parties and uses the example of noise pollution. A simple answer would be to use technology to reduce or avoid polluting nature. Nozick cannot see that scenario since in his view everything equates to cost/benefit in terms of property owners. That's dangerous because it enables polluters. Even at the time he's writing, it's erroneous and short-sighted to think one can enumerate the victims of pollution when the reality is that all organisms on land, sea, and in the air suffer and perish. In his theory, prohibiting those who engage in risky business should be compensated for their "disadvantage" (1974, p. 83), but that benefits logging companies and cattle ranchers. Some businesses without a good report card regarding nature actually receive government subsidies. There should be prohibitions and fines where the money goes toward forest reclamation and conservation, not back to rich corporations and billionaire executives.

Nozick asks what device in the state of nature is an enforcement mechanism (1974, p. 89). The answer is natural selection. There are, to use his language, prohibitions of enforcement in nature. He focuses on individual rights, but in nature, which he evokes repeatedly, it's not the human construction of rights per se but forces of individual or group selection, genetic mutation, or drift. In his theory, contrary to this argument, there is no state of Gaia but only a collection of individuals. However, according to Rawls (1999), whom he cites (1974, p. 90), these individuals across species cooperate in symbiotic mutual aid to everyone's advantage with restraint and yet liberty. That sounds like a state, but Nozick sees binding enforcement on individuals. Indeed, so it is in a forest and yet a success of fairness. He talks negatively about the emergence of group rights inhibiting individuals (1974, p. 90), but we see successful individual-group activity in social animals from ants to primates where, in line with Walzer (1983), there are different values placed on parts of resources to ensure fair distribution.

In a discussion of fairness, Nozick seems to advocate for the cheaters and outliers (1974, p. 93). He fails to see that in nature there are mutual obligations accounting for the self-regulation of Gaia. Only in Nozick's conception of human nature are there instances of numerous disgruntled rebels and cheats, for in nature they have mostly been selected out since it's not always beneficial for survival and reproduction. It's precisely because Nozick lives in a Rawlsian free, welfare state that he can imagine individual liberties that do not enhance the status quo. He does not think fairness should be enforced (1974, p. 95). This claim is debatable in the argument for forest sovereignty. Energy companies do not need to continue mining or drilling as much as they want, producing and selling to their benefit fossil fuels that pollute the air all organisms breathe, especially when there are non-polluting alternative energy sources. Where is the fairness in that, or in deforestation for logging or creating massive livestock ranches that pollute and deprive forest wildlife of their habitats. Intimated by Walzer (1983), maybe these are the social goods the political community wants to distribute; but the meaning of community values can change.

As for prior exchange agreements to which Nozick refers (1974, p. 95), there was little to no representation of local people, Indigenous people, or the forest flora and fauna. In an anthropocentric manner, his theory about the state of nature focuses only on modern humans. He goes on to say that individuals have rights to know how they'll be treated by any dominant agency (1974, p. 102), but no such process happens with how humans treat forests. Instead, it's all about human-centered property and financial exchange. Nozick insists that individuals have a right to resist anything they deem unjust and unfair, but there's no mechanism to enable forest dwellers to resist. Since the world is centered on human needs and desires, whether in peace or war, there is no protection for nature. Following the reasoning of Nozick (1974, p. 106), what humans perpetuate upon nature is unjust punishment for no wrongdoing. Again, following his line of thinking (1974, p. 108), humans have applied a monopoly of artificial force against defenseless nature, as seen in corporate industrial animal farming.

Nozick does talk about protecting others (1974, p. 110), but not surprisingly animals are not included. He says it would be of no purpose to define population or territory size in determining a state (1974, p. 117), and such is the case with sprawling Gaia, he'd have to admit. In nature, he says, enforceable state rights are minimally those of the individual (1974, p. 118). If so, humans cannot deny the political existence of Gaia.

We know there is power in nature evident from volcanic eruptions and earthquakes to tsunamis and hurricanes, but the human machinery to tame nature is a monopolized imbalance, as even Nozick might say (1974, p. 121). Nature does not attack humans, as a wolf does not attack prey. Rather, the wolf exhibits its evolved feeding behavior in a biotic system just as the energies in nature often erupt in geological systems. Randomness in nature has nonetheless created ordered movements. Humans, on the other hand, methodically attack nature and disrupt feeding balances in deforestation. In terms of defense, using Nozick's language, humans should not build weak structures on sand, near oceans, or on hills susceptible of crumbling. Nature simply is in gradual shifting configurations with occasional surges, and that's not news. Humans, however, have helped unleash these natural energies by their own acts of deforestation that have weakened the typically stable self-regulation of nature and the atmosphere.

Rather than say, as in the past, humans must defend against nature, one would argue for cooperation with the state of nature. There could be recognition that most human practices are the ultimate causes of destruction, whether building tall towers on fault lines or constructing housing developments in a water-deprived desert or river flood plain. Nozick goes on to talk about compensation and punishment (1974, p. 135) in his conception of the state of nature but is not correct since his natural state is fully human. Since humans have upset the balance, there might be forest restitution. Indeed, following Locke (*ST*, secs. 7–9) Nozick uses language like "moral boundaries" to suggest everyone has a natural right to punish (1974, p. 137). He also uses the word "wrongdoing," but humans have cleansed their actions of any faults. If there's only one punisher, by Nozick's thinking, that's a *de facto* state (1974, p. 140). By that logic, with the human state against nature, in just fairness and due compensation human nations should remedy their harms to the environment. In his logic, Nozick argues for a minimal state, but that's unrealistic if humans across continents in many nations are the sole punishers of nature. There is no way to minimize or justify how severely humans have unfairly harmed nature and have unjustly penalized people at risk of flooding, superstorms, heat waves, etc.

Concerning distribution, Nozick says there's no central locus that holds power to control resources (1974, p. 149). That discounts the obvious wellspring of resources in nature, from which humans freely draw. In short, here's my claim in one sentence. In stability, nature allocates resources consistently among forest inhabitants; humans take unfairly from nature and do not distribute evenly. Nozick might agree with me, and he goes on to discuss property acquisition and transfer along with a sense of entitlements. Where we differ is in how he sees raw materials and land for the taking by the first newcomer. That land and those resources belong to Gaia and Indigenous people. Nozick recognizes unjust acquisition and illegal transfers, but only among modern humans. He disfavors distributive justice since it slices up resources according to needs, a systematic and stilted pattern of weighted sums (1974, p. 157). Rather, following F. A. Hayek, he favors distribution according to the value a person generates for others (1974, p. 158), supposedly trickling down benefits.

Production and distribution are not separate; producers are entitled to their goods as even Nozick says (1974, p. 160), but like other thinkers he refuses to see this for forest animals. In this case, the forest is the one who bestows the most advantage to humans though many people often abuse the gifts, unlike other organisms who indeed contribute to Gaia. Modern humans, assuming entitlements, engage in willful, unjust distribution of forest resources that belong to wildlife who sustainably maintain them until irretrievably taken. In his theory, socialists interfere (1974, p. 163) with the libertarian entrepreneurs, but the opposite is true with entrepreneurial capitalists interfering with the ecological balance of Gaia's social commune. Nonetheless, like humans, forest animals have a property right in their persons and labors (1974, p. 171), Nozick would have to admit, not just humans. He says only in distributive patterns people are justified in having property rights over animals (1974, p. 172), but that's not so with consumerist, industrialized nations employing his entitlement theory for forest resource acquisition and transfer.

If, as per Locke whose thinking Nozick seems to follow, labor adds value, then corporations who use low-wage people and big machines to mow down forests are entitled to own the land. This conclusion is not solid. Well before any humans, there were many forest organisms tending, seeding, and tilling the earth, the main point and a major claim for sovereignty in the forest state of Gaia (Brehm, et al. 2019; Mortelliti & Brehm 2020; Estrada-Villegas, et al. 2022). Nozick admits there are questions surrounding exact measurements in any value-added property theory, such as how appropriation can be detrimental to others. He goes on, however, to employ Locke's notion that with whatever is taken there should be enough "left in common for others" (1974, p. 175; Locke, *ST* ch. 5, sec. 27). That guarantee does not occur with forest animals who suffer great loss of habitat, possible death, and probable species extinction. I've already discussed this in the section on Locke, but it's interesting to note that in spite of historical time and what happened with the commons in England to the loss and detriment of the working poor, Nozick still advocates for this stance. He insists that private enterprise will enhance a property's productivity for the social good far better than multiple owners (1974, p. 177). This, too, is false, since corporate ownership of vast forests and other land generates pollutions that harm all earth's inhabitants. As is typical with Nozick, he lives in a realm of abstract ideas and equations, not reality, for even in his own time destruction of the Amazonian forests was a social concern. He claims that Locke's *proviso* of enough left for the common good will not be fouled by the free-market system (1974, p. 182). That's false if one looks at the extent of deforestation for capitalist gain.

Nozick goes on to claim that in Rawls' theory, social cooperation creates problems of distributive justice (1974, p. 185). This might be so in human societies, but for the most part it's not so with forest organisms across species. Groups of the same species would, of course, vie for similar resources. In this respect, there's a conundrum. Is the forest *polis* closer to Rawls with equilibrium of cooperative sharing and Walzer with distribution of goods based on value; or nearer to Nozick where individuals obtain goods through their own means and are alone justly entitled to them? We come back to the baffling question about the nature of justice. Nozick presents as a problem for Rawls' liberal theory the idea that too many people in social cooperation will have equal claims on any "joint product" (1974, p. 186). First that's not the case in the forest with different resources for different species in different niches in line with thinking from Walzer. Besides, some resources, niches, and certainly habitats are shared where one species is unaware of another's particular needs in line with thinking from Rawls. Second, all humans have an equal claim to the clean air and water generated from all types of forests. Libertarians cannot claim air as their own because it has passed through their lungs. The vision with Nozick and other political theorists is a myopic focus on human labor and divisions thereof (some artificial) in market exchanges and trading terms, which are only pieces of earth's ecological jigsaw and typically damaging to the environment.

In Nozick's world there are incentives and entitlements for personal gain, but in nature, not to oversimplify, there are motives to avoid suffering and achieve care for self, kin, and others in the group, if there is one. There are checks and balances in nature selecting out behaviors and genes that don't add to the survival and especially reproductive success of species. This process of descent with modification is essentially union among species and environment, a far cry from the lament of Nozick who complains that the Rawlsian justice as fairness doctrine, or Walzer's differing value-based distribution, leaves some privileged humans with less than they really desire (1974, p. 194). Forest flora and fauna use their natural assets and abilities to serve others locally and globally with their ecosystem engineering in feeding, defecation, physical movements among others, and decaying bodies. Does that fit with Rawls, Walzer, or Nozick? Rather, the point is to show the feebleness of quasi-Enlightenment political theory if left to ignore nonhuman entities. Rawls would say everyone can in common lay claim to natural assets whereas Nozick would counter by asserting that only individuals are entitled to the benefits and property acquired through their abilities (1974, p. 228). Consider, according to this discussion of forests and earth's biome, which of these perspectives benefits the world overall.

Nozick criticizes Rawls for assuming that equality is inherent in a theory of social justice (1974, p. 233). While we see differences within and across species, resources are distributed and shared according to those differences, as Walzer might say. The net sum is balanced, but we have to, following biology, see predation as a feeding behavior and a means to stabilize populations. As Plato suggests, justice is written in the overall community and not in any individual (Hadas 1950, p. 138). Presumably, Nozick would not be comfortable with the fact that a forest society manages to arrange evenhandedly the activities of its inhabitants. Nozick looks not only to allocations and distributions but also to the basic origin of services, meaning those individuals entitled to share their skills (1974, p. 235). Once again Nozick examines Western industrialized society without taking a worldview or especially a glimpse at the natural world from which humans evolved. Darwin recognized what he calls the struggle for existence, but there's mostly in nature an equal opportunity of access to resources so that all individuals have a chance to survive, reproduce, and adapt. Some don't, but that's a condition of nature and not an academic question of justice. Where Nozick embraces variation in competition to support his ideas, Rawls does not. Either way, their theories are rooted in human societies separated from the natural world.

A state of Gaia is somewhere between these two seeming extremes of political thought where there is entitlement through adapted talents along with equal opportunity for nature's bounty. Nozick discounts people's rights to goods and services because others have created those products. If that's so, then humans need to recognize how forest organisms own the rights to the "substructure," as he denotes it (1974, p. 238), of all resources there. Forest animals and plants have a vested interest to maintain the ecosystem for themselves and their descendants. Humans, apparently, do not hold that view or take that position, but perhaps the anti-animal and abuse-of-nature attitudes will change with a new generation whose very existence is threatened by notions of entitlement and exceptionalism.

As workers, flora and fauna take risks to invest in their habitats and are compensated, but then consuming humans, who have made no such contributions to Gaia, steal that fruit of labor without any just recompense. According to Nozick's thinking, especially in the section "Marxian Exploitation" of

Chapter 8, the inhabitants of Gaia are the workers and producers and yet exploited. He taunts social democrats to take the risk of starting a business to produce goods, and that's what Gaia does, but then humans take whatever they want without remunerating the owners. This stilted attitude goes against his philosophy. Forest organisms share but humans do not because of anthropocentric ideas about the value of humanity written in Western philosophy and the lesser worth of "animals." Nozick, too, is good at this, making a dig at "animals" while promoting anthropocentrism (1974, p. 243), ironically in a discussion of envy and self-esteem where he places humans as better than other organisms. In truth, each species has evolved differently, not in a hierarchy or teleology. In fact, let's argue against strict demarcations between species, the whole point of Darwin's work. We are a community of life forms, and humans could learn to live together cooperatively with nature and not act against Gaia in aggressive competition. There are hints of that thinking in parts of some countries.

The value of forest workers' products means more to them and a huge array of organisms in a bio-network than to the desires of some consumerist humans. More so, the production value of forests benefits the air and water supply for all of earth's creatures, and that includes humans. So contrary to Nozick, Marx was correct in the following formulation. Gaia is exploited by capitalist-consumer societies. Gaia's inhabitants have the skills as workers to keep forests healthy, but rather than receiving protection from humans they are killed or forced to extinction as their habitat homelands are plundered and reduced in size. Nozick ignores all of this with his laser focus on libertarian privileges ignoring the "socially necessary" (1974, p. 258) innovative labor Gaia tallies for earthlings. Humans have assumed the resources of forests have no utility for the workers there and value only for themselves, but science proves otherwise.

Nozick claims that individuals have a right to speak about what affects them (1974, p. 270), but he will not grant that right to forests since, as a libertarian, he'd deem them as human property capitalists exploit for profit. He claims that those economically advantaged are more politically inclined in a "nonminimal state" since they can in that arena exert more influence (1974, p. 272). Yet we see power dominance from corporations and rich nations operating in developing countries, especially regarding forest lands, what Nozick might term minimal states. He further claims that such power brokering could be the result of illegitimate politics, but that's not necessarily so. We see legitimate governments and their politicians, including those in Western nations, privileging a range of deforestation for profit from cattle ranches in Brazil to palm oil and sugarcane plantations in Borneo. Not surprisingly, the real problem is that deforestation and unsustainable land use has become culturally ignored by the general public and accepted by those in monetary power. Nozick looks at all of this as ways of strengthening the state, which he opposes but for property, personal, and financial protection. The culprits of deforestation, though, are people: local people, politicians, executives, etc. The very individual liberties and freedoms Nozick espouses are those that could be culturally modified for a sustainable, ethical, and healthful future.

Even Locke, whom Nozick invokes, would say that humans don't own forests since they did not produce them under careful control (1974, p. 288). The next sentence is worth repeating. Forests are naturally organic and not manufactured industrial warehouses. Unlike a human-created machine which needs intervention to keep functioning, forests are self-regulating and sustaining, as they have been for many millennia. Nozick ends his book with a discussion of an anthropocentric utopia (1974, p. 313) whether urban or agricultural, a nowhere land. Instead, it's more useful to speak of the evolved world, damages to it, how it can be healed, and how we can educate people away from unethical and unsustainable practices like disproportionate consumerism, excessive meat eating, and wastefulness that lead to deforestation. Some of that change is happening, so it's not a nowhere land. Finally, for reasons that seem contrary to his philosophy, Nozick mentions Peter Kropotkin (1974, p. 310), who demonstrates mutual aid among species in a socially fair community. Nozick's utopia is the minimal state where individuals are accorded dignity, rights, and are not used as a means for resources to another's ends (1974, p. 334). That notion is precisely how humans should treat Gaia but hitherto have mostly not.

Utopia: Degrowth for Forest Growth

Giorgos Kallis and Hug March (2015) imagine movement away from capitalistic growth to a socioecological environmentalism in degrowth where limitations are not scarcity but encourage climate utopia. Degrowth is a form of decolonization when production is not capitalistic but based on local economies scaled with equitable distributions. Currently, industrial humans are ruled by private property, nation-states, capital and money, armed forces, economic institutions, etc. None of these exists among the flora and fauna of Gaia, and yet it's a state of self-sustaining and self-determining communal plenty more often than not if left large. Like other animals, we have a consuming instinct; but whereas they forage for food, search for mates, and build temporary shelters, we are acquisitive for material goods generated by deforestation, oil drilling, mineral mining, corporate agriculture, industrial production, etc. Perhaps human societies could reverse capitalism, as Kallis and March suggest. The point is that there's already a pattern in nature worth emulating, a model which should be left alone for climate stability.

Additionally, in this regard, consider Indigenous people who on a global scale respect and protect biodiversity through their small communities in relationships with nature. As the U.N. suggests, by virtue of their sustainable practices Indigenous people should be granted "forest tenure rights" (Ibrahim, et al. 2025). Why can't those rights be arranged for wild plants and animals, too? Importantly, degrowth is not regression but a confrontation with the faults of current enterprise. Coming back, again, to Locke, the issue is to know when *how much* is *enough*; the problem is that capitalists don't want limitations or collective sharing of the commons manifested in forest flora, fauna, and fungi where there is equal access to resources in spite of competition. As Kallis and March (2015) show, scarcity is a social creation; for sure, scarcity among forest plants and animals is compounded by loss of territory, pollution, climate degradation, etc., forced upon them by modern people.

Karl Marx would likely embrace degrowth; Robert Nozick would be skeptical; Peter Kropotkin would nod in agreement since the process of change would involve mutual aid seen already in nature and among Indigenous people. Beyond Rawls and Nozick, and flowing from the notion of degrowth, the distribution of social goods according to Michael Walzer (1983) differs according to groups and their size; within a common welfare, some resources and products can have differing meanings and value, a pluralistic approach that favors social justice. This picture works in a natural landscape and among Indigenous people, but some industrial societies struggle to accommodate communitarianism and political ethics.

Social Goods and Spheres of Justice

From the concerns of Nozick, let's segue to broader issues to understand how these middle chapters are central to the argument for forest sovereignty. Michael Walzer (1983) goes further than Rawls and answers Nozick by saying that equality comes in the form of recognizing physical being along with emotional and mental states. With our knowledge of ecosystems, we can apply that standard to forest wildlife. Let them live, thrive, and die according to natural laws. Like Rawls and Nozick, Walzer does not speak of the natural world or wildlife. For him the focus is on humans. The problem of equality is not just differences (e.g., rich/poor) but how one group imposes domination over another to create inequality that festers resentment or a feeling of "subordination" (Walzer 1983, p. xiii). We don't know that animals or plants feel subordinate to humans, but for many centuries humans have espoused injustice toward nature through domination of ecosystems, subordination of species, and repression of individual mobility in an established habitat. Walzer writes about spheres of justice and distributions with no mention of wildlife, as outlined in this discussion. In fact,

Walzer (1983, p. 65) inaccurately speaks of the "malevolence of nature" with his focus on human needs.

Walzer (1983) talks about "social goods" like membership, territory, security, welfare, and work. Social goods can have different meanings across groups based on interests and preferences. We find these distinctions in nature because of adaptation, and all the pieces seem to fit. Evidenced from the discussion so far, wildlife is more than capable of inculcating and sharing social goods. Humans alone did not create a distributive community. As has been shown, in nature material goods are passed around among many others in a complex distribution, as Walzer might say. He'd go on to suggest how there's no real unity, no single distributive system, but rather the sharing of social goods through pluralism and heterogeneity in cultural diversity. We see that in wild nature. Not all social goods are the same, and so they are distributed differently among various species in a variety of ecological niches in shared communities across biomes in the biosphere. Walzer might see the sharing of social goods as transactional among humans; in the wild, there are relationships. Distributions in one area, no matter how small, affect outcomes in another place, contingent or not. Nevertheless, distribution can create conflict in both the human and natural realms; the distinction is in how forest flora, fauna, and fungi have evolved adaptations, ecological niches, and populations to accommodate different needs. Billions of humans still vie for many of the same resources based on perceived needs.

A primary social good for Walzer (1983, p. 31) is membership: with whom, to whom, and from whom. There are many instances of membership in forest flora and fauna, from flocking and pairing geese and beaver environments to plant associations. Distributive justice occurs among members and excludes others separated by real or artificial boundaries. That too one can find in nature. Territory is another social good and, for the most part, a central concern of this book. Any theory of justice according to Walzer (1983, p. 44) specifies the rights of state inhabitants. Since Gaia is not recognized as a state by politically powerful humans, wild flora and fauna have been denied rights in spite of their mutual interdependence as ecosystem engineers providing, in that respect, a massive social good for all forms of life. Security and welfare also come up as social goods, and here Walzer (1983, p. 64) addresses how a community provides for everyone in mutual cooperation. This we see through natural adaptations in addition to, more prevalent among humans, cultural evolution. Work, too, is a social good according to Walzer, and among the forest flora, fauna, and fungi their employment in provisioning for themselves provides feedback assistance to many other organisms. The foraging of one animal opens areas of food for another; the blossoming of plants encouraged by animal browsing creates pathways of nourishment for others.

In a nutshell, we find in the natural world, still not perfected in some human nations, a social contract to distribute resources among members and across populations according to, as Walzer would say, needs (1983, p. 82). This bonding surpasses "differences" in a theory of justice that recognizes, in line with Darwin's ideas of variation and divergence, sovereignty where property ownership is discouraged and social goods are distributed without bounds. Sovereignty as state power regulates social goods and places boundaries around how the state distributes justice (Walzer 1983, p. 281). In nature, there's distributive justice through the evolution of ecological niches.

Moving forward, let's expand the political philosophy of this and the previous chapter in the next three sections.

Participatory and Representative Democracy in the Forest

The social contract for Locke is, as Jonathan Wolff says (2006 p. 39), a "voluntaristic obligation." A question revolves around who in the forest has given themselves over to the power of humans. Rather, humans seize power over forests and animals in the name of humanity. The human state compels one into submission to its ways and laws. Forest creatures have not granted or consented to human state authority over them expressly or tacitly. Worse, the human state with logging, mining, industrial agriculture, housing developments, etc., offers no benefit to plants and animals in those ecosystems. So, without any advantages or consent, human state rule and exploitation is not justified. Locke would say one has a natural right to freedom, especially since no animal has agreed to human rule. As Schmitt (2009) points out, there is no single freedom but an array. Corporations and industrial agriculture do not have *carte blanche* freedoms to rule forest lands. Their freedom should be limited and harmonized with the rights and liberties of the vital lands they usurp with epidemic destruction. Corporate "freedoms" unregulated and unchecked by governments are a threat to a healthy planet. Supposedly, the Enlightenment social contract theory is what elevated people from a state of nature to modern societies, but no such legal agreement came into place in such a way. Even so, given change over time a social contract from ten generations ago should not be absolutely binding on present people or with people imposing any such contract on nature. On the other hand, forest creatures engage in a "participatory democracy" (Wolff 2006, ch. 3) where all are active in the biology of the ecosystem. Each one has a rightful and consensual interest in the local ecology.

By definition, there's very little freedom in a society where people are not permitted to participate in making rules that govern them. In the forest, through countless ages of evolutionary adaptations, flora and fauna have established their own guidelines of behavior. Rules of the forest worked for animals until humans began to tamper with ecosystems. It's as if humans are authoritarian dictators over the livelihoods and freedoms of animals. Many democratic nations condemn dictators who abuse human rights, yet there is little concern among most politicians for the liberties and rights of forests. Conservation of land, water, and air is a concern for all. Resources don't have to be sold off by national leaders to corporate bidders since they'd not want any other nation appropriating their territories by force. Starting with Plato, say Will Kymlicka and Sue Donaldson (2014), political citizenry focused on self-rule grounded in reason, self-control, and moral reflection, as if animals are incapable of their versions of those virtues. Animals have sets of norms and grasp the subjective experiences of others, too (Andrews, et al. 2024; Westra, et al. 2024). As Schmitt (2009, p. 195) argues relying on Plato, a government should be run by experts. This begs the question as to why humans believe they can do better than the self-regulating system of Gaia that has successfully managed forests, wetlands, and grasslands for millennia. Expert humans are not able custodians of the earth, as has been proved over the past few hundred years by their industrial pollution and deforestation. Exceptions of course are the small bands of Indigenous people who, like the flora, fauna, and fungi of forests, work sustainably in synergy within bio-networks as well as some districts of progressive nations.

Some people might say only expert, informed opinion can resolve questions about sustainability, sovereignty, and ethics in the forest. So far, political and corporate experts have damaged forest areas vital to plants, animals, and the atmosphere. Instead, with evidence based on eons of a self-sustaining and healthy system, inhabitants of the forest are the experts through their ecological behavior on environmental issues. Yet they have been mostly ignored by many politicians, corporations, and people who have voting powers. Scientists might tell another story. Unlike humans, forest animals know what they need, and generally the processes of foraging and feeding aid their health. Humans, however, can eat or drink to excess, drive cars crazily, or take unnecessary risks across many social or financial behaviors. Schmitt (2009, p. 207) quotes Edmund Burke who says that a government represents the interests of one nation. In terms of a forest, this statement can be dissected as follows. There is one focus of Gaia, which is to maintain sustainably the well-being of all flora, fauna, and fungi. On the other hand, the great forest nation of Gaia consists of many different species' interests and ecological niches that constitute a single home. Modern representative governments recognize regional rights and yet typically have national interests, in a similar way. It's as if there is absentee rule by powerful elites lobbied by corporations governing the forest, where human and not animal interests prevail.

Looking more closely, we might be able to label forest law as participatory democracy. There is active involvement in decision-making about rules to which one is subject. Take as one instance a social insect like the honey bee. Carol Gigliotti (2022) provides evidence of research attesting to bee democracy. This is a given since social behaviors are time-tested evolved adaptations. In modern, human societies participatory democracy is not practical since large groups dominate. That could happen in the forest across pack, troop, or herd groups, too. In the forest, generally, there's no agenda as one finds in industrial and even developing nations, for example evident in which political party holds power in the struggles for dominance. Since participatory democracy would consume all the time of everyone, citizens instead elect people who legislate and administer laws in a representative democracy. The problem for industrial societies with far reaches into forests is that powerful groups (e.g., logging companies and corporate agriculture) lobby those representatives in their favor. The will of the people might not be acknowledged. Elements of representative democracy exist in species that have alpha leaders (gorillas) and close allies (chimpanzees). In these examples it's less about managing public affairs and more to controlling rule; still, participation by others functions here. In line with Mill, it's fair to say forest creatures are independent and self-protective, seen among female elephants, key to their vitality and survival. We don't have Plato's wise guardians in the forest, nor are they required. As in a political system, so too in nature according to Darwinian thought there are checks and balances, a dispersal of powers, with an interest to sustainably maintain the environment in spite of self-interest.

Can this be proved? Indeed, if one looks to the forests before human intervention the evidence is there. Mill was reluctant to give voice to poor laborers fearing they'd create an unstable economy by upsetting the distribution of wealth (Wolff 2006, p. 98). Whether this prediction would be true or not, the point is that forest organisms have a good sense, conscious or not, that their needs are tied up with the interests of others in their locales. For the most part, in the forest government of non-interference there is no underrepresented minority since all have an equal vote regarding the life of the state. Their freedoms and liberty are typically reduced or eliminated, as in their deaths, by human invasion that upsets the natural order.

The Transnational State of Gaia

In line with my argument, Susanna Hecht, et al. (2016) favor more investment in tree and forest management and less in fanciful technological solutions. The means for carbon capture and storage are already in nature, and humans might want to recognize the benefits of additional woodlands, even in urban areas.

The longer we wait, the more the climate crisis will cook forests to the point that we won't be able to restore them or engage in reforestation or rewilding. Living forests are an obvious solution to fighting climate change and should be our prime focus. Technology can be useful but becomes old, rusty, obsolete, and hence must be junked and replaced. Solar panels are reliable and possess longevity, but their lifespans pale in comparison to some tree species that can live several hundred or even up to one thousand years. Gaia supports an energy balance that can become unnaturally disrupted by human activity. One study (X. Song, et al. 2018) shows net tree cover gain in subtropics, temperate, and boreal climate zones with a loss in the tropics. This study measures only a sliver of surface matter in a much larger and deeper environmental problem.

Such studies do not reveal how much tree cover has been lost over hundreds of years because of human development or how much tree capacity is needed to restore the equanimity of Gaia. There seems to be a shifting baseline syndrome where we will accept less forest cover since that's all we know. None of us was around many hundreds of years ago when trees covered entire continents. According to William Ruddiman (2003), deforestation began with early agriculture in Eurasia at around 8kya and irrigation for rice at 5kya. Abandoned farmlands naturally rewilded during plagues in the Middle Ages might account for the Little Ice Age (1300 to 1900), he says. This strongly hints that more tree cover will run as a counterbalance to slow global warming, though we are already in a crisis mode.

So far, I've been making a distinction between human nation-states, or those with governments prone to securing borders and resources, and the natural forest state of Gaia. Moving past political philosophers yet borrowing from ideas in *The Green State* by Robyn Eckersley (2004), let's take this a step further and illustrate Gaia not just as a forest state but as a transnational state. This is where the political philosophy of yesterday is crucial for tomorrow. We already have a global green state in nature, but its dominion is unacknowledged.

Eckersley (2004) asks how a nation-state should manage its ecology. The implied answer is there doesn't have to be wholesale resource depletion only for economic gain (Nozick) but some consideration about maintaining a stable environment for the future (Rawls). Rather than fostering a liberal democracy, she favors an ecological democracy. This means that a green state within an environmental movement would be less concerned with its nation-first sovereignty (Nozick), especially in competition with other countries, and more focused on revising political norms to create a healthy global environment through cooperative action (Rawls). Environmental and economic problems are now global with few borders. Gaia is the primary or transnational "state" in which all other national cultural identities with fabricated borders are located. In this way, Gaia needs to be respected as much, and perhaps more than, the fences, stone walls, and metal barriers erected by one group of humans to keep out another group. As Walzer (1983) would say, forest inhabitants across continents know the meaning and value of their social goods and how to distribute them for mutual benefit. That's a healthful transnational and not an economic transactional approach.

Nonetheless, Eckersley goes on to acknowledge that state actions have environmental consequences, sometimes good but mostly bad. The upshot, given the limitations and yet wide range of states, is that she envisions a *transformation* of "state power" (2004, p. 6). Politics is power struggle, so conflict within a nation and between states is not easy to reconcile ecologically. Even with sovereignty, some state borders are not recognized, and nations are attacked. Eckersley privileges legitimizing the democracy of states; we could do so, then, for Gaia. Just because a state has power to log, mine, or deforest biodiverse areas for development does not mean it should do so; and if it does, under our current global climate crisis let's seek accountability. Loaning from Val Plumwood (2001), Eckersley acknowledges how environmental injustice means that the actions of a few removed from ecosystems they harm benefit politically and economically at a cost to poor people and local flora and fauna. We too often rationalize our unecological ethics. Mick Smith (2011, p. 193) sees "biopolitical dangers" in greening nation-states; for him, sovereignty and biopolitics equate to the governmental control of nature. Yet, a transnational state of Gaia seems plausible given this century's political will to grant forest sovereignty so that such territories can be preserved.

Moving forward, concerns both social and environmental are considered together in an eco-centric approach that benefits all forms of life. Who will be the "good international citizens," asks Eckersley (2004, p. 13). Could we find them blended in aspects of thinking by Hobbes, Locke, Rousseau, Smith, Mill, Marx, Rawls, Nozick, or Walzer? With a new theory, we want to visualize a sovereign Gaia, which left alone manages itself transnationally barring environmental catastrophe thrust upon nature by humans. Challenges for a human green state, Eckersley seems to indicate, include nation-first mentalities that promote competition, global capitalism with only a mindset of profit for the powerful already rich, and liberalism in democracy that might overlook prioritizing solutions to ecological problems (2004, p. 14). All three are interdependent. Based on what we see over history, but for some recent developments, one wonders if any state will even commit to full sustainability given anxieties about border defense (Hobbes), economic growth (Smith), human class structures (Marx), political power shifts (Rawls, Nozick, and Walzer). With nation-states predominantly concerned about balancing military and economic powers, there's less incentive to declare forest sovereignty.

National apprehension about security is especially true given how natural resources, whether coal, oil, timber, minerals, water, etc., are used to fuel a state's economy or military. On the other hand, consider the forest state of Gaia as a sustainability model where resources are not depleted but cycled around in nearly perpetual self-regulatory maintenance unless disturbed. Nation-states have a rational incentive (Locke) and could be motivated to forest preservation for national and global good (Rousseau), looking at the situation from an anthropocentric perspective. Increasingly, social order in some states might begin to become unstable as yet more citizens call for progressive reforms to the climate crisis. Activists and green thinkers, employing cultural evolution, might tap into social components of climate anxiety to swell networks of active voices. Unless quashed by conservatives and capitalists, this process can begin anywhere from grade school, high school, or in college. Even some enlightened politicians are beginning to see how environmental issues left unresolved pose threats to national and economic security. Worse, the climate crisis is beginning to envelop some areas in real-time disasters that can't be stopped. One question is whether nation-states will cooperate more fully to avoid mutual destruction. Not to be cynical, but some nation-first mentalities might revel in the thought of a political rival suffering severe consequences from climate change. In the U.S. we see the two opposition parties openly stifling each other's agendas. Mutual cooperation is difficult to find. State borders are artificial, so if one state falls from climate disaster, others are soon to follow. States in the imminent water crisis within the Western portion of the U.S. and on other continents continue to bicker but must seek cooperative action.

The problem with sovereignty is that one nation-state cannot dictate how another uses or abuses natural resources within its borders, even if misuse results in deforestation, soil erosion, water depletion, air pollution, etc. The paradox is that by granting sovereignty to the forest system of Gaia, then no one nation-state can violate its dominion of flora, fauna, fungi, waters, and lands. Through development, whether a human state is socialist or authoritarian, it will pollute, and infected air and spoiled water will spread to other states. Foul elements are not the rule with the transnational forest state of Gaia and why, as a huge carbon sink and air/water purifier, we might want to leave it alone, not to suggest human states can continue unbridled pollution elsewhere. While human nation-states are making some positive movement toward small levels of sustainability and climate control, the forest state is the true transformative domain in fighting and neutralizing much of the climate crisis and why we might want to negotiate its sovereign immunity from human intrusion. With business as usual, developed and developing countries will proceed along capitalist trajectories where the bottom line is accumulation of wealth that chokes liberal pockets calling for ecological preservation.

Just as nation-state agencies might work together toward forest expansion, preservation, and sovereignty, states might strive for ecological harmony across the board. The challenge these days is not simply to address symptoms or invest in potential technology that might someday alleviate the climate crisis. Collective action from an informed public working on local levels to pressure nation-states to act on forest, wetland, and grassland expansion and preservation is needed. Technology, rather, should be geared to greening cities from the underground, street level, and upward in buildings that rely on renewable energy, recycled materials, and carbon capture for any small amounts of fossil fuels they burn for heat. The global economy need not be a competitive race to see who is the dirtiest but who is the cleanest. Dumbing down existing forest preservation laws for developed or developing nation-states will only escalate the climate crisis as well as problems related to sourcing clean air and water. Eckersley (2004, p. 68) talks about corporations operating as "pollution havens" in countries with weak environmental laws, empowering companies to produce and export many goods at little cost to increase profits. Germany and the Netherlands have good environmental regulations and strong economies, so environmental deregulation is not necessarily an ingredient to corporate success but surely one for environmental disaster. Uniform environmental policies discouraging deforestation across nation-states would be helpful. When nations make fighting the climate crisis profitable for businesses, there's an incentive to act. Of course, the motive could be more virtuous and about preserving forest ecosystems to help foster "societal transformation," as Eckersley says (2004, p. 71), in tune with ideas of this discussion.

Modern nation-states have a role in creating and preserving the transnational forests of Gaia as components spurring technological, economic, and social progress in the fight for climate control and justice. Gambling planet earth's future for powerful, rich politicians and executives is unfair. There could be some degree of public enforcement over the ecosystem by not abdicating control of forests to profit makers, as Eckersley (2004) would agree. Mine is not an argument to destroy capitalism; rather, in forest preservation there are new frontiers to invest capital, such as rebuilding old cities across the globe to be less dirty and prevent deployment of population expansion to further disembowel forest areas. An entire labor force can be employed in reforestation and rewilding. Liberal democracies with multiple (competitive) parties are more prone to social and economic change than totalitarian states, so open societies would help encourage environmental activism. Eckersley (2004, p. 87) questions these assumptions since self-interests prevail, evident in the slow progress of environmental action across capitalistic, liberal democracies. While there are legal and administrative policies and practices in place to manage a state's daily affairs, often there's no coherent plan for attenuating environmental risks.

On the local level, modernizing cities using new outlooks to optimum green capacity can take pressure away from destabilizing forests, wetlands, or grasslands. With such modernization technological innovation will precede and yet follow new policy initiatives to include rooftop or vertical farming of veggie food, solar or wind panels for energy, and rainwater capture and reuse so that forest edges and cores can be spared decimation of those resources. The possibilities for greening urban communities are endless, from parking lots that capture water to buildings that capture carbon, from green gardens in pocket parks to recycling and reusing centers within walking distance. One simple solution would be to let biopatina (fungi, moss, lichens, etc.) grow on building exteriors to help capture carbon. Cryptogamic covers like cyanobacteria, algae, fungi, lichens, and bryophytes act as ecosystem engineers in carbon and nitrogen cycles (Elbert, et al. 2012). Old city-owned buildings can be repurposed exclusively for green uses, like urban vegetable or wildflower pollinating gardens. The will of administrators must be able to prioritize and accomplish such goals driven by public demand.

As Eckersley says (2004, p. 105), echoed in other parts of this book, some ecologists recognize how human social and moral spheres include "ecological communities" of wildlife. There's no human/nature dualism. Humans are not excluded from the natural realm as many climate catastrophes are proving. Whereas industrial humans have been instruments of the climate crisis, they can mitigate many of the problems by implementing forests to capture carbon, clean air, and hydrate water. These positive actions do not exempt nations from curbing polluting effects or from working on other climate control mechanisms, like promoting a vegan culture and embracing renewable energy resources. Some authors (Feigin, et al. 2023) argue that moving to a plant-based diet is the only way forward in terms of tackling the climate crisis. Eckersley (2004, p. 111) offers an "ambit claim" for a democracy centered in ecology. Anyone affected by the risk of climate change should be able to participate in corrective decisions. By the same token, no forest flora, fauna, or fungi have agreed to or welcomed risk and damage to their habitats. Better representation is required in an open dialogue, or as she says, a "democracy *for* the affected" (2004, p. 112). Rejecting this argument, some will say that nonhuman beings are not rational and cannot communicate with us, so we can do as we wish. This anthropocentric thinking ignores basic biology. All living organisms strive to live and control functions in their habitats, often on intelligent and conscious levels.

Harms are not only to humans, nor local, nor confined to a specific time. Deforestation affects multitudes of organisms and ruins air and water quality for life forms over the globe now and in the future, so it's difficult to justify on economic grounds alone. While nature is a reality, Eckersley (2004, p. 121–22) suggests that it's also an objectified "shifting social construction." Groups might see nature through their status, class, and nationality differently. Reporting on forest devastation or loss is insufficient, so most people are not informed and cannot effectively evaluate causes, effects, and predictions. Real efforts establishing environmental communities, councils, etc., have helped the preservation cause, but imagine how much moral power such organizations, governmental or not, would have with conditional mandates by multi-border nation-states for forest sovereignty. Biologists, ecologists, bioethicists, and activists won't have to compete with commercial interests or lobbyists. This means that those promoting and preserving forest sovereignty won't necessarily be working in or for the government but with and for the global public (Eckersley 2004, p. 142). Moreover, Eckersley hints (2004, pp. 161, 164) how there should be a public voice of moral opposition and influence that is more accepting and wider than narrow program politics. Beyond any cultural or national community there could be more multinational thinking focused on global ecology. At present, we seem divided, discordant, and nibbling only at the fringes, but some progress has been made. As a preliminary step, it would be helpful for governing parties across nations, whether developed or developing, to cultivate a culture of forest ecology with shared goals in social and political discourse.

We already have many international agreements, as Eckersley (2004) discusses. For instance, take the 1972 Stockholm Declaration, principle 21, which permits states to abuse resources on their territories as long as there's no damage to other states. This policy echoes what we saw in Locke and Nozick. Of course, this might have worked 50 years ago but is obsolete now. No human nation-state needs to have sovereignty over earth's interconnected bio-network. The U.N. Charter 2.4 says no state can forcefully interfere with or encroach upon another's territory. Again, this is outdated considering the dirty air from burning forests for beef cattle ranches or palm oil plantations that wafts across continents, or the plastics from countries that accumulate in the bodies of whales, or the acids that destroy coral ecosystems. The 1992 Rio Declaration on Environment and Development permits countries to develop their economies as needed with nods to environmental considerations. Let's endorse new rules that are binding and not merely declarations. Some might cry Western imperialism, so then let the new forest order spring from Indonesia, Africa, or South America. Nation-states don't need a blank check as possessors of exploitable forests but could be party to agreements overseeing them as responsible caretakers of a globally shared borderless nature. The collective wisdom of Indigenous people as good stewards of Gaia needs to be included in any policy discussions and enactment.

Ecology and Ideology

For Nozick, liberty and equality are not compatible, mainly where there is some form of equitable distribution. On the other hand, similar to Rawls, Ronald Dworkin sees liberty in an equality of resource distribution (Farrelly 2004, p. 73). Dworkin calls for a distribution of wealth under legal orders. We already see such fair distributions of resources in nature since plants and organisms consume and use resources differently because of evolved body size, function, life history, and habitat adaptations, comparable to thinking by Walzer. Gaia's arrangements permit organisms small and large to evolve a self by making choices, both conscious and instinctual, in their circumstances. Dworkin says equality among citizens means respecting them as persons and not leaving their welfare to arbitrary social or market forces (Christman 2002, p. 80). With the emphasis on person, we see anthropocentrism. In the forest as opposed to cities, all animals have an opportunity for a good outcome. Though that's not always achieved, the starting point is usually not one of deprivation, poverty, discrimination, etc. This is not really utilitarian but an equalization of welfare. In human societies, there's little equality among costs and benefits across citizenry where some are deprived of resources while others are wasteful. In the wild, assuming extensive lands, there's more access to shared resources from which one can responsibly plan a life and thrive with little harm to the welfare of others. If, as some argue, humans are an exceptional breed endowed with super rationality, consider the irresponsible choices they've made individually, politically, and socially and the costs to nature.

For example, financial ties between environmental groups and big oil and gas corporations help forest-destructive companies amass "reputation insurance" and praise (Hari 2020). There are blatant conflicts of interest between lobbyists and environmentally unfriendly corporations, and if this unethical practice continues, there's little hope to save or expand forests. Technological proficiency cannot ever fully restore the true value of nature, not what is deemed useful for a healthy planet (Elliot 1982). We seem to be willing to compromise our ethics by accepting falsified or artificial versions of nature from misinformed media and deceptive corporations. Environmental concerns could dig deeper than sugar-coating the effects of pollution via corporate campaigns to win public support (e.g., big oil spills). There are moral issues regarding biodiversity, soil systems, wildlife, Indigenous populations, and biosphere egalitarianism where all organisms flourish and not only rich and powerful people (Naess 1973). Greater species diversity optimizes food webs overall and evolutionary adaptations within populations (Hutchinson 1959), so it makes no sense financially or ideologically to destroy transnational forest biotic ecology that supports all forms of life.

Liberalism has "market fixations" regarding every aspect of our lives, from food consumption and marketable goods to government interference by any number of agencies, suggests Timothy Luke (1999, p. 11). These multiple emphases have ultimately had negative effects on the environment and have removed people from understanding the importance of nature in their lives. We are becoming subject to fluxes in nature, which market systems in developed as well as developing nations have caused. Removing the barrier of corporate dominance over nature we are able to see our lives and forest life in a new way. Citizens of the world have ceded control of their lives to political-corporate brainwashers, which tend not to be environmentally conscious. As Luke says, nature is treated as "dead matter" and an "alien entity," (1999, p. 51). Nature is not dead, and so Luke (1999, p. 22) argues for an active "ecological populism" on a local level independent of corporate or political-administrative control. This strategy can work in some parts of the world where people manage forest edges (not cores) with sustainable farming. That's already happening in parts of Indonesia. While the political and corporate machinations of nation-states were the causes of many environmental problems leading to our climate crisis, those devils can also become the angels of change. Most of the world's population has a relationship with commodities (e.g., cell phones, cars) when we could somehow establish and maintain a relationship with the natural world.

Inspired by Luke's (1999, p. 61) discussion of permaculture, or self-sufficient green systems I've repeatedly noted in Gaia, here are a few ideas that cross political theories, although there's a Marxist slant. Energy production from renewable sources could outpace usage, with the ability of storage and affordable distribution. Discourage new development, focusing instead on greening existing urban areas. Any development could preserve and maximize soil integrity and water use. The retooling of urban areas might focus on local needs and concerns with an eye to linking suburban vegetable farms (not meat or dairy) nearby or within buildings. Preserve existing nature at all costs to maintain native wildlife habitats and biodiversity. Pollinator gardens could be created and managed by young people in cities and suburban developments. Keeping in line with a green revolution, Luke (1999, p. 76) talks about the business of corporations fashioning ideas of class to sell an "ephemeral culture" of expensive and disposable material goods (e.g., luxury automobiles) to a few hundred million people in cities across Asia, Europe, and North America while many others live in poverty. An ideology of buying and selling does not embrace permaculture. In other words, value is not placed on life, seen in nature's renewable resources to vivify earth, but on possessing symbolic notions of status.

Whereas most people on the low end of the income spectrum reuse, repurpose, and recycle materials because of financial necessity, those with higher expendable income levels discard and replace material goods regularly based on desires. Luke's (1999, p. 105) point seems to be that many people worldwide have become technology dependent rather than community oriented. Widespread technologies, even by the end of the 1900s, led to deforestation and desertification, greenhouse gases and other air pollutants, ruination of arable lands and fresh water with chemical fertilizers, etc. These practices are not ecologically regenerative. Luke points out (1999, p. 106), for instance, that capitalist structures dictate how much more energy is wastefully produced even if it's not really needed in some sectors of the economy. Capitalistic systems and not necessarily communities decide how to use technology, often to the disadvantage of nature and sometimes to local people.

On a similar note, nation-states who exercise control over resources distributed worldwide maximize their economic advantage and leverage over people and the natural environment. Luke (1999, p. 157) discusses some authors with the notion that perhaps modern humans are meant to industrialize nature in co-determination. That explanation only justifies the wants of rich people at the expense of nature. Luke cites (1999, p. 201) and seems to support economists who advocate for a morality in economics that excludes violence, but that viciousness is all we've seen in most nation-states, developed or not, with deforestation and non-optimal economies that promote organic decline. Any social ecology, to use his phrase, must avoid domination of forest areas that are capable of self-management. That's most of Gaia and why, like E. O. Wilson, we should advocate for expansion and protection of more forest lands. Just as we should not force onto the poor the high cost of wealthy people's standards of living, it is equally unethical to expect forest flora, fauna, and fungi to die as they feed the exorbitant wants of rich nation-states.

Chapter 4

Reflections on Political Theory and Other Considerations

Now that we've established ideas from primary texts by Hobbes, Locke, Rousseau, Smith, Mill, and Marx and have covered modern interpretations from books by Rawls, Nozick, and Walzer, it's time to reflect broadly on these thoughts and look at more culturally inflected notions about "humanity." This chapter is a bridge to the conclusion. First, let's say a few words about T. M. Scanlon (1982, p. 110) who, making a nod to Rawls, is opposed to utilitarianism, but who emphasizes a theory of contractualism or "informed agreement." For Scanlon, there's an interpersonal morality of giving to others what they are owed concerning problematic distribution. Scanlon leans toward cooperative living and recognizing the equality of others through relations without agonizing about right and wrong. The key to Scanlon seems to reside in how contractualism offers values none could reasonably reject (1982, p. 111). This relates to the argument here since forest wildlife would reject arguments from politicians and executives to harm them. The "rational desire" (Scanlon 1982, pp. 118–19) of the politician or executive contradicts forest well-being, where wildlife will not accept self-interested wishes of financial gain for powerful people.

The Social Contract

Both Hobbes and Locke believe that governments were created by people to overcome the hardships and limitations of nature, but look how well Gaia operates. Social contracts are useful but artificial. For Locke, government secures the right to unlimited private property for the bourgeoise (S. Smith 2012, p. 170), hence industry and consequently Adam Smith's *Wealth of Nations*. Large areas of private property could, however, be kept biodiverse in exchange for tax reductions. Disapproval of monetized private property when it's anthropocentric and not eco-centric is echoed by Peter Burdon (2015) who espouses an earth jurisprudence. Moreover, there's tension between Locke's property rights and Rawls' equitable and Walzer's pluralistic distribution. What's needed is a social contract that legitimizes the rights and liberties of Gaia's "citizen" inhabitants. In line with Scanlon (1982), how could forests or environmentalists like E. O. Wilson refuse that offer. For Hobbes, peace is forced through a powerful ruler; for Locke, there is communal peace through legislation of power (Schmitt 2009).

Not everyone agrees to all terms in an Enlightenment social contract, and animal interests are not fully represented. Human states and corporations want to raid forests to log and mine; animals do not agree by their pained responses to such incursions. Rather than self-interest, for justice Rawls asks for "impartiality" (Wolff 2006, p. 154). This is not to say that animals are not without interests, for they exhibit concern for survival and reproduction in many ways. However, their interests are often confined to certain ecological niches in a locale, what Walzer might call their spheres of value. Humans, on the other hand, have grand self-interests that extend beyond any physical or hypothetical borderlines. In their limited self-concerns and refined habitats, animals are more impartial observers and not takers of other life, excepting feeding adaptations to aid metabolic processes. Humans can choose not to eat animals, evident in many ethical vegans. Animals, in some respects, don't necessarily know what other species need or desire, so they leave them alone as if others are free.

Conversely, humans know the limits of the forest and the needs of the inhabitants but do not act with impartial justice. Generally, they act on self-interests. There's no "veil of ignorance" in line with Rawls (Wolff 2006, p. 155), for corporations are motivated to profit from nature, whereas animals are disinterested from exploiting resources for profit. Mary Anne Warren (2001) says that Rawls limits moral rights to humans, but of course many social animals, whether wolves or primates, demonstrate moral agency from instinctual kin care to group norms (Monsó, et al. 2018). We can only wonder, then, who is more just in this social contract. It's clear human groups artificially force scarcity of land and the draining of resources on wildlife. There once was abundance in nature but no more, no fault to the animal managers. For Rawls, justice is resource regulation in society between abundance and scarcity (Wolff 2006, p. 157), but he fails to consider the transnational forest state of Gaia in this equation.

There are principles of justice for Rawls: equal rights to liberty with social and economic goods for those most in need (Wolff 2006, pp. 157–58). These principles are based on freedom, differences, and opportunity in that order. By attacking forests, these pillars of justice fall. Raiding forests is to the benefit of many people; yet, goods produced from natural resources are not readily shared with the poor. Inequity is not necessary if it engorges consumers from industrialized nations while adversely affecting the climate because of deforestation. There's the added problem of harms and death imposed on forest creatures who maintain the health of the flora. Political philosophers, psychologists, and economists can talk of Rawls' "difference principle" or equal distribution (Wolff 2006, p. 158), but they have in mind only humans who benefit from the labor and lives of plants and animals.

Rawls does not favor utilitarianism since he sees inequalities in a welfare state dependent on a free economy. Rawls says the underprivileged can reject utilitarian inequalities that disfavor them (Kymlicka 2002, p. 55). First, why can't everyone have access to basic liberties? Second, those who are unfairly disadvantaged could have priority to social and economic benefits not by excluding anyone but gained through cooperation (Kymlicka 2002, pp. 56–7). In the forest there is equal opportunity where circumstances fit the ecological niche. Individual success depends less on arbitrary circumstances and more on efforts. This is not so in human societies where many are disadvantaged by physical or mental handicaps, not just adverse circumstances limiting choice and abilities. In the forest, Darwinian selection and Malthusian economics would sift out those who cannot survive or successfully reproduce. What's left are prosperous individuals and species working together, sometimes at odds, to ensure the viability of resources to the benefit of all according to distributions across ecological niches.

In nature there's no weakness in social contract theory. Rather than a free-for-all with one pitted against another, there's a set of interconnecting paraphernalia. The authority of Gaia cares for the workers who, in turn, keep the property functioning. Social contract theorists and their critics assume there's no moral obligation among creatures in nature. Since morality is *mores* or the ethics of interpersonal behavior, there's a contract where none in nature is morally above or below another. The forest is not anarchy because for the most part Gaia protects inhabitants from scarcity and uncertainty (Kymlicka 2002, p. 62), which is not always true of human nations who engage in civil, cultural, or state wars. Civil disobedience, or a revolt against a government that is not preserving peace or protecting property, is espoused by thinkers like Locke and then Thomas Jefferson, says Schmitt (2009, pp. 102–3). Forest creatures cannot rebel on their own behalf against governments who willfully harm them.

A truly legitimate government representing all its inhabitants and not powerful groups or rich corporations alone would value forests for their own sake and also for the cleansing effect they have on the atmosphere. Non-violent protest envisioned by Henry David Thoreau worked for Gandhi and Martin Luther King, Jr., but has not yet been fully successful for those representing forests. Those of the forests have been excluded from the political process and must, then, rely on the sincerity of proxies to advocate for them. This is no small issue. Like many immigrants, animals are perceived as foreign threats, as pests, or are simply ignored until humans want their bodies or lands. Unlike people, many animals cannot easily integrate into human society, though authors like Sue Donaldson and Will Kymlicka have made that argument for some species in *Zoopolis* (2011). Other than fringe areas where human and animal contact is inevitable, and which should be peaceful, it's best to let forests alone unless salutary intervention is required due to human degradations.

So, it seems that Rawls, following Locke, Rousseau, and Kant (Kymlicka 2002, p. 62), could be speaking of forest creatures when he talks of fair equality in spite of species differences. Concessions for survival and flourishing have been made through ecological adaptations. Because most humans desire or need the same resources, there is often less concession and more inequity, why Walzer devised spheres of justice based on the meaning and value of different social goods. While there is, according to Rawls, a "veil of ignorance" in society where individuals are unaware of the talents or finances of another (Kymlicka 2002, p. 63), we don't need that veil in nature. The veil is supposed to guarantee equal divisions, which occur naturally in a forest through adaptation of niche construction. In nature, ignorance regarding one's place, control of a habitat, and the whereabouts of food, offspring, kin, and others is not psychologically advisable. Among forest creatures it's not even possible to talk of contracts, differences, and inequalities since they all work according to their evolved needs through descent with modification. Much of these shared resources and equal divisions presume there's enough land for grazing, browsing, foraging, nesting, traveling, etc., exactly the point of this book. No single organism lays claim to forest soils, trees, or waters.

Libertarianism and Moral Liberty

A nation-state might not have an obligation about an individual's outcome but should provide basic rights of freedom to choose an appropriate path. There's a question of how one shapes a life and what social or market forces harm or support her. For libertarians, property is not a given to be equally distributed but an earned good. This self-interested acquisitive behavior is in stark contrast to the free flow of resources in nature for trees, plants, soil, microorganisms, and mammals. Possessive ideology is what has destroyed nature and its inborn ability to regulate air, soil, and water for all. A libertarian could argue that under the current climate crisis, a privileged few should own water. That capitalistic tendency and even the libertarian call of property ownership without leaving enough behind violates the natural rights of others and victimizes forest workers. Animals not only use but create resources cyclically. In a paradox for libertarianism, wildlife enjoys but yet does not possess forest property.

Kymlicka (2002) asks who engineers the social state. Without a central government in the forest nation, there's a free exchange of goods affording equal entitlement to resources without restriction. This thought goes further than Nozick because there is no power system controlling property ownership. Yet Gaia is efficient and fair in its distributions and freedoms. There's a moral equality in line with Kant and Rawls that stems from self-governance and self-ownership (Kymlicka 2002, p. 107). One could say that because some animals eat others they are treated as a means and not as selves. This is true only from the human perspective since feeding behaviors evolved with metabolism. In contrast to most human food systems, wildlife feeding replenishes the forest ecology. Some animals, like big cats, are obligate carnivores. Many primates, including humans, are omnivores. Their food is not removed from the forest. The salient point is that the transnational state of Gaia as one organism with multiple parts does not consume itself into extinction. Compare that to how human systems can hypothetically lead to extinction, some species of which are already suffering early demise. Even in the human body there are microbes and parasites working with or against each other in our guts. Rather, people are the ones who treat the forest and its animals and plants as instrumental resources or means for their own ends of profit as pleasure. This practice is not surprising since humans have mistreated animals for many thousands of years as beasts of burden or as food.

Likewise, in a utilitarian system, Rawls would complain, some few are used for the advantage of many others (Kymlicka 2002, p. 108). This is an enigma since the forest appears utilitarian and yet acknowledges the rights and liberties of individuals. With authors like Wolff and Kymlicka there's overlap of political theories in looking at how organisms manage the forest. One size does not fit all, as modern politicians try to force their ideologies on large portions of society. For instance, where Rawls would see rights to resources Nozick would proclaim individual rights. We have both in Gaia, with exceptions. Contrary to libertarian philosophy, Walzer would say that social goods like talents, products, and harvests are shared according to value, as in the seed, stem, and rind dispersal of fruits by primates, which in turn feed creatures below the tree canopy to encourage growth diversity. Like liberal egalitarians, to go further, most forest animals are not enslaved or owned by another, so there is freedom to create and own the fruits of one's productive talents. Exceptions to this can be found, for example, in some ant species. Each forest organism has a distinct method of foraging or is a means of becoming food ensuring survival for others and hence continuation of the systems in Gaia. Humans have generally upset those structures with their linear thinking in terms of commercial enterprises.

There are individual rights and responsibilities in a community of mutual aid. There is self-ownership while others can own some of the results of another's labor. In many respects, each is a resource of the other in the state of Gaia, a somewhat liberal egalitarianism and the secret of its long success. Networks in forests are not quite redistribution of (Rawls), nor are they unrestricted access to (Nozick), resources, but yet there is self-ownership and equality within the governance of Gaia since different values of social goods are understood (Walzer). Whichever political theory one uses, an assumption is that forest lands are for human use, distribution, or ownership without acknowledging that animals are the rightful heirs to those properties. In a real sense, the flora, fauna, and fungi all combined have created the forest and so have legitimate title to the physical property and inherent properties there. It's technically wrong in examining these political philosophies, when discussing property rights, for Kymlicka to say, "no one created the land" (2002, p. 110). Some people would say the will of a God is the creator; others would say the energies of soils, microorganisms, plants, and animals of Gaia are earth's progenitors.

The human assumption is that there's no land transfer if there was no original human acquisition per Nozick (Kymlicka 2002, p. 111). In fairness, Kymlicka says that whatever property is now held in human hands privately, originated in nature. Since lands and resources were taken by force at some time, then it only stands to reason that there should be an equitable distribution by governments and corporations. Indigenous people were and still are better stewards of forests than executive men and women in expensive suits sitting in fancy corporate boardrooms. In any accounting, plants and animals must be granted what is rightfully theirs as originators where their decomposed bodies literally created the natural resources exploited by humans. Even Nozick takes issue with properties (e.g., from workers) seized by force (Kymlicka 2002, p. 112). As Kymlicka reports, in seventeenth-century England the enclosure of common lands began and Locke, as noted, justified this exclusion of others by saying it was legitimate if one took only what was needed and left some for others.

This enclosure had actually started earlier and ramped up by the eighteenth century when dissipated aristocrats began to envision their lands as profitable farms. Taking only what's needed and leaving enough for others is the practice of hunter-gatherers and especially animals but not the business plan of corporations or capitalists. The human justification for enclosure is to prevent many interlopers by transferring control to a few. First, animals control lands and so forests, including any wetlands or grasslands, should be enclosed for them. It's conceivable even in one view that the enclosure can be governed by humans, who must be environmentalists, local people, ethical philosophers and the like. Second, forest creatures when undisturbed never deplete the commons. Invasive destruction came because of humans. If Nozick (Kymlicka 2002, p. 115) thinks land should be enclosed as per Locke's outdated notion to avoid depletion, then the forests should revert untouched to the native people, flora, and fauna who have ably maintained it as earth's air purifier, water hydration system, and carbon sink.

Garrett Hardin (1968) has some insights about the tragedy of the commons where fair and equitable land use was spoiled by cheaters. He suggests minimizing reliance on technological solutions and instead looking to make changes in moral values concerning overpopulation and the impossible task of maximizing the greatest good for the largest number on a limited tract of land. It's not mathematically possible since land, resources, etc., are by the laws of physics limited, seen in how U.S. cattle ranchers take advantage of federal grazing rights pummeling grasslands to erosion. Forests, grasslands, wetlands, and similar natural spaces should not be commons for human exploitations or waste dumping. There are natural population checks in forest lands, but human populations remain unchecked and have increased so dramatically that demands on common lands are unsustainable. If this destructive land use policy was previously known, along with the carbon dioxide emission studies cited earlier, it's disturbing that moral values have not changed and have produced a tipping point of no return toward climate catastrophe. Activism on a local level must be embedded within the state, which has monetary power and influence over any hypothetical global commons to address issues of deforestation and climate change (Mansbridge 2013).

Creating markets to buy and sell resources like water harms small farmers and forest people and is a tragedy of the commons, but Paul Trawick (2003) sees a viable alternative among Indigenous Andeans. Water markets appeal to some economists and legislators who are often motivated by financial concerns. This issue of the commons raises the question about who owns and who can control a resource as fundamental as water. Keep in mind that Gaia's systems, especially rainforests and other woodlands, are responsible for much of earth's atmospheric and land moisture, ultimately managed on the ground level by plants, animals, and Indigenous or rural people. Sustainable solutions by small groups contradicts Hardin's (1968) dilemma of the commons becoming depleted of resources. The tragedy, however, as argued here, is the loss of the commons from local people and animals when appropriated by corporations and political bureaucrats. Trawick shows how among the Andeans and across Latin America large, economic interests are the culprits, not rural people, diminishing the commons by privatizing and selling resources. His solution is to transfer water management to local communities, much in line with how I see climate resource management granted to forests, wetlands, and grasslands. Trawick's proposal is previewed by Bård A. Berg (1996) who sees how the Norwegian government interfered with reindeer pasturing by Indigenous Sámi, hence creating a tragedy of the commons that did not previously exist. The tragedy, as mirrored here, is when capitalistic human economies establish property regimes that disrupt how plants, animals, and native populations sustainably care for the commons. J.

Terrence McCabe (1990) likewise notes how use of land among Kenyan Turkana pastoralists is sustainable and not destructive; this too supports the view that the tragedy of the commons is not caused by flora, fauna, fungi, or local people but by political and corporate mishandling of natural resources.

A large part of the problem in these theories of political philosophy is that the focus is on human capital, property, ownership, and transfer. These thinkers use words like "resources," but they are not speaking as ecologists or ethicists on behalf of animal conservation. We see how many business people somewhat ridiculously refer to commerce as an ecosystem. Political philosophers also talk of self-determination as if that applies only to humans. Look to the forests and see deliberation, reflection, and intelligent thinking in many social species ranging from wolves or elephants to corvids and great apes. Individual animals know, like humans on an instinctual and cognitive level, what is good for them. How does one claim social animals have no meaning in their lives, especially when their sociality is rooted in the forest as ecosystem engineers. Humans have treated subjectively aware forest animals as chattel when they destroy their habitats and thus remove the ability for self-determination. Instead, inhabitants of forests can be mutually advantageous to human cities and rural areas (e.g., carbon reduction) were people to cease deforestation and begin a worldwide effort of rewilding.

Forest plants and animals need as much freedom and land as possible, more than they now have. We do not have to enter into a contract with the forests. Instead, human states and nations could establish a convention for non-interference with collective action for expansive tree preservation and planting. We don't simply need a cosmetic fix of pocket or corridor forests; large core areas of untouched trees, lands that might also include wetlands and grasslands, would be preferable. Animals and hunter-gatherers in large forests pursue self-interests, sometimes without constraint, but because of scale there are no tremendous losses but gains or at least a stable state. While not explicit, morality is evident in forest labor for mutual advantage because of high rates of reciprocal productivity (e.g., seed dispersal, biomass growth, etc.). Utilitarianism, liberalism, and libertarianism are all concerned with individual free agency since on some level freedom promotes production (Kymlicka 2002, p. 139). Humans alone do not have greater moral liberty than wildlife to satisfy consumerist desires when the forests are more beneficial to the interests of lives within and outside of their borders because of carbon sequestering.

Property Rights and Distribution

Realistically, forest lands are not for public or private use; they exist because of evolved adaptations by many species of flora and fauna, as well as microorganisms, working together over eons to construct a collected series of habitats and ecosystems. Like mythological Adam and Eve before their expulsion from Eden for polluting that garden, creatures of the forests have maintained and preserved the lands well. That high degree of care cannot be claimed by modern humans, including some local people in developing nations who can inadvertently spoil environments. At the same time, animal rights activists, say Kymlicka and Donaldson (2015), working in modernized cities have a cultural mindset at odds with Indigenous people living off the land and their ethics governing human-animal relations where they hunt to eat. The Indigenous people don't necessarily see animals as things, or objects in the way of development, or pests, as might nations favoring commercial use of forests. There are problems here, nonetheless. For example, Indigenous people will hunt whales or primates without any real necessity in a modern world that can provide high protein plant foods. Indigenous people don't have a birthright to kill off other living organisms any more than outsiders have the right to take their lands and resources. Cal Flyn (2022, p. 46) strongly suggests with evidence that some nineteenth-century Native American tribes over-hunted and depleted wildlife. Indigenous inhabitants of forests, like the animals there, do however tend to use and share resources for all. Western "civilizing" religious conversion of some Indigenous people has led to a colonizing of their lands and resources over hundreds of years, whereas previously the small bands of local people mostly controlled their properties sustainably.

How does one reconcile animal rights and Indigenous rights? How much hunting is enough? Is hunting, according to my line of thinking, of a different order in that the dead animal is not food for other forest creatures and organisms? The significance of a wolf feeding on another mammal, or birds feeding from the remains, or microbes feeding from the carcass, or the bones filtering into the soil as nutrients is critical to the maintenance and survival of woodland property. Human hunters do not quite contribute to forest sustainability in spite of any so-called rights they might have, and Kymlicka and Donaldson (2015) question if aboriginal rights supersede those of wildlife. Who here is developing land economically: the forest population or the humans? Is there moral conflict in my argument that seems to coerce all people into what I call good environmental behavior? To answer that question, Flora Lu Holt (2005) sees Indigenous people in a Catch-22, as she calls it. Opposed to my view of leaving vast tracts of nature to wild, she advocates conservation tied to social processes connected to politics and resources. Nomadic, low-density populations of Indigenous people, in many areas, are gone; also, traditional customs and practices now employ some modern technology (e.g., guns and market economies). Conservation biologists might question the romantic notion, given Western influences, of the harmony between Indigenous people and nature. Protectionist enforcement by conservation biologists might not be the answer; rather, as I advocate in the proposal with modern societies, Holt too thinks public awareness and cultural evolution among Indigenous people about ecology are key components to biodiversity preservation. At any rate, as I continue to stress, the only true conservationists are the naturally occurring flora and fauna in forests, wetlands, and grasslands.

The 1992 U.N. Conference on the Environment and Development in Rio de Janeiro discussed climate change and biodiversity protections but failed to reach agreement on forest sovereignty since nation-states asserted their financial and resource rights as primary. While there were approvals to forests as part of a "global economy," many nation-states shrugged off financial support for any venture to fund a transnational state of Gaia (McGinley, et al. 2012). Since 1992 many Latin American countries, more than U.S. states, have come to see the value in sustainable forest management by holding inventory, protecting boundaries, and preserving at-risk species and riparian buffer zones. Laws are "social, ethical, and political creations" says Kendra Coulter (2023, p. 97), and progress regarding the legal defense of animals appears in Colombia with laws that recognize how chordate and vertebrate animals share sentience with humans and are consequently afforded protections. Many Latin American countries, rich in biodiversity, have increased educational initiatives and legislative protections for wildlife, Coulter goes on to say.

Indigenous people in Andean Ecuador, Peru, Bolivia, and northern Chile revere Pachamama or Mother Earth in their acknowledgment of the rights of nature. Rani-Henrik Andersson, et al. (2021) point out how Ecuador in 2008 amended its constitution to affirm nature's right to exist; in 2010 Bolivia passed a similar change to remove human dominance over nature; in 2014, New Zealand made Te Urewera land a legal entity and no longer a Crown possession. Indigenous rights and reforestation are also key components. Considering all that has changed since 1992, and with financial support from rich Western institutions, the time to reconsider forest sovereignty has arrived. Modern humans, in contrast to forest wildlife and Indigenous people, plunder resources for the benefit of the politically powerful or for corporations satisfying consumers. Governments cannot truly justify taking lands from Indigenous people. More to the point argued here, animal advocates and forest conservationists justify rights equal to or perhaps greater than those of human inhabitants in Gaia.

In addition to other rights enumerated, the right for a healthy climate alone justifies forest sovereignty. Animal freedoms in large forest areas are totally connected to human freedom. Without climate change mitigation many people, not only those in developing nations, will lose the freedom to live and travel because of roads and neighborhoods lost to fires, floods, mud slides, or desertification. According to Schmitt (2009, p. 123), rights of ownership help dispel conflict. Forest dwellers, with that logic, must be superior beings since the evolution of their ecological niches, not property ownership, is what has helped them develop adaptations of conflict avoidance. As Schmitt goes on (2009, p. 124), Plato advocates non-possession of property to eliminate divisive tensions and foster unity. The sense of belonging all together exists among forest dwellers, and yet they are called beasts or savages. As they struggle for existence in diminished territories, often at the hands of industrialized and even developing nations, there can still be a state of contentedness when a habitat replenishes food for foraging. This condition of satisfaction has been lost in modern people, Rousseau thought, since human societies have created false divisions and inequalities (Schmitt 2009, p. 126).

Intellectual property is also a consideration. The forest state of Gaia is at once a vast expanse of material, from microbes and worms to lichens, old-growth trees, and mega fauna. While biological mechanisms are at work, there's also an intangible element about this self-regulating system acknowledged by Lovelock (1979) that nearly infuses all of the physical stuff with something intellectual, borrowing from the Latin etymology of that word. In synergies, elements of the bio-network understand how to operate in unison. When humans touch forests, they are not only violating the physical goods and freedoms of Gaia's inhabitants, but they are also damaging intellectual property. Humans do not share but usurp these various properties. We know that many animals have consciousness (Kohda, et al. 2019) as well as emotions and feelings, so it's not far-fetched to imagine a collective consciousness in Gaia. The clean air and water generated through the conscious and unconscious communal work of forest Gaia ought to be free to circulate for the benefit of all living creatures.

Humans who limit forest growth prevent this sharing of a common good. Schmitt (2009, p. 155) declares as "obvious" that nature "belongs" to humans. Large areas of forest require protection, creating massive areas of nature belonging to forest individuals and not to be privatized or patented by unscrupulous entrepreneurs. Every element of the forest is a nonhuman invention owned by Gaia not to be infringed upon by invaders. Excluding animals from use of the forest they have helped evolve and consistently cultivate is an ethical error. Large multinational agricultural corporations genetically modify plants, gain patents for them, and then enter into lawsuits against anyone who misuses the plants, even if the seeds blow onto someone else's farm unwittingly. By that logic, all forest flora is owned by the fauna and microorganisms who encouraged those plants to evolve as they have. If natural resources are wealth, as nations believe, a small group of powerful people or executives cannot justifiably claim possession from the owners or users in schemes of unfair distribution. Deforesting huge tracts or removing resources impoverishes those inhabitants, and once they are suffering from extreme habitat loss, governmental and capitalistic operations are to blame.

Free Markets

Robert Costanza, et al. (1997) discuss ecological economics, but only insofar as ecology impacts human interests. The authors point to a range of problems: unsustainable population growth exceeding earth's carrying capacity; technology that depletes resources and pollutes; land conversion for corporate animal agriculture, logging, development, etc., that erode soils and harm biodiversity. In the decades since those alarms were raised, the balance has always fallen to corporate profits and not environmental responsibility. In terms of the world's material economy, resources should be sustainably preserved and shared cyclically so that ecology and economics are wed. Humans disproportionately appropriate most of earth's biomass. Add to that depletion pollutants in the biosphere, whether carbon dioxide or nuclear waste and fallout (Stäger, et al. 2023). Costanza and his co-authors assert that their findings do not equate to an ecological economy. As they note, the academic discipline of economics arose in the middle of the nineteenth century from moral philosophy, a center since lost. Society's ills and remedies, institutional responsibilities, financial markets, and research sciences are drivers of health and wealth, common goods for humans but not for forest plants or wildlife. For much of the late nineteenth and all of the twentieth centuries, nation-state economies were concerned only with monetary growth at the expense of nature. Sustainable finance and investment in a circular material economy is discussed by Maryna Vahabava, et al. (2022). This means that production and use of natural resources is not an economy separate from nature but part of biodiversity preservation and climate crisis mitigation. Nature can be regenerated by eliminating material waste and pollution and by recirculating products better and more widely.

Asserting the greatest good for the largest number, or maximizing utility to enable welfare for many, utilitarianism is the backdrop against which all other political theories seem to work (Kymlicka 2002). Utilitarianism is more to social structure and not individual responsibility. Utilitarianism might remedy Locke with the human use and transfer of land for the greater good, but it's not for the welfare of all living organisms. Rather, this is about mercantilism, industrial trade, and the inheritance of wealth only for small groups, like corporations or rich families. Free-market capitalism does not benefit all living creatures, just as it does not help all people. On the other side, there are economies where the state is in control, as with forest organisms and animals. Here, the goal is not necessarily profit but satisfying population needs in the forest where resources are strictly allocated in ecological niches. In the forest there seems to be a combination of freedom and statehood, an evolved economy where there is coordination, profit, and benefits for all. There's no waste but only balance and equity, not counting natural or human-made disasters.

There's no real market motive among forest inhabitants though there are supply-demand issues. For millennia, these have been balanced out via Darwinian selection and Malthusian economics. As in the free market, there is forest equilibrium of resources per animal, generally speaking, given enough land. This balance gets thrown off kilter with human incursion and initiates a negative chain reaction with loss and then extinctions. The supply-demand of the forest is controlled by transnational Gaia unless disrupted, so all creatures have food, water, and shelter. This is not a Platonic ideal but a reality. Shortages and poor quality can result naturally or, more likely at present, from human-induced climate change. In that case, forest dwellers adapt or rely on fallback foods. Part of this state-of-the-forest control is adaptation by natural selection, an equalizer. The problem with a human free-market economy is that in the drive for profit, production can outstrip resources. In spite of Adam Smith's *Wealth of Nations* (Wolff 2006, p. 146), there's no benevolence bestowed on either system, free market or Gaia's economy, but forests naturally evolved to replenish resources in sustainable cycles.

In a human state-planned economy there's more negative control over consumption with inflation, distribution bottlenecks, etc. That's not the case in nature where foods can be, in a best-case scenario, plentiful for all without restriction on growth or consumption. In a sense, forest resources have been seen by humans as public goods, and so they've acted as pathological free riders ransacking the state of Gaia. In consequence, along with deforestation and the diminishment of feeding ranges, abundant earth has been reduced. The costs of this rampaging could be paid by human states and returned to the forest; if not, forests together with Indigenous people should be left alone. While freedom-loving capitalistic markets can thrive, they can also falter, as is often the case seen in their unsustainable seismic rumbles with mergers, buyouts, bankruptcies, lawsuits, recessions, etc.

There's less waste and more equal employment in the forest economy. The question is which system is overall more efficient and for whom. In capitalism, workers are oppressed to engross the profits of business owners, but that's not so with the interconnecting mutual aid economies of the forest where there is no exploitation, generally. There are, then, far fewer disproportions in the forest state economy than the capitalist free-market economy that exploits forest resources including animal lives. In a capitalist enterprise, shareholders do no work but profit. In the forest, all inhabitants work and profit, an equality not seen in capitalism. Capitalist free markets do not really protect their workers or the forests. Animals are not capitalists since they don't completely control the distribution of goods; yet, the productions of their labor are for the advantage and well-being of others.

Communitarians and Distributive Justice

Humans disagree on how goods and equality should be distributed in a society (Rawls and egalitarians, including Walzer) if at all (Nozick and libertarians). Material conditions can determine the opportunities one has to achieve the same goals as those privileged with better chances for success (Marx). Proportions appear in nature (e.g., body size, function, and life history) with the result that organisms obtain, for the most part as evolution proves, just amounts of space and resources to survive successfully, reproduce, and adapt. Wild animals know how to live freely, and their choices and actions in the forest serve the ecosystem and atmosphere. Left to their own devices in the forest, industrialized humans, unlike hunter-gatherers, have made bad choices, like deforestation for logging or corporate agricultural. In some ways, the evolved behaviors and cultures of forest animals derive from the communities in which they live. It is beneficial for all that forest creatures have inherited the cultural operations, like ranging and feeding techniques, from their community. The values of Gaia matter to all forest organisms. This community ethos need not cut against liberalism. Cultural norms even among forest animals can be positively modified if they benefit the group, like Japanese macaque monkey potato washing or African chimpanzee nut cracking. Importantly, any change should come naturally, as an adaptation in nature, and not forced on Gaia by human hands.

Communitarianism wants to maintain what is seen as the communal "moral fabric" across a plurality of groups and their goods (Farrelly 2004, p. 98). Communitarians are less for individual liberalism and more to social traditions, with a sense of community at the core. For the communitarian, liberal individualists can become disconnected from the community and social behaviors. This is exaggerated. Besides, any social group, human or animal, balances a mix of individuals, especially exceptional ones, with conformists. My view does not deny the importance of community but questions its values and beliefs. We can't persist holding repressive customs, like patriarchy, if they are detrimental to a free and thriving community. Ethics permits a revaluation of morality. Some customs and traditions can weigh too heavily on and constrain reform. Communitarians turn away from individual rights and freedom in liberal theory or the self-determination of thinkers who seem to ignore the impacts on a community (Kymlicka 2002, p. 212). Unlike Marx, who argues for no state, no private property, and no religion, communitarians favor homogeneous religious preferences in traditions for a common human good, implying less individual freedom, especially for out groups.

The emphasis on a common good is partly evident in the forest though each species evolved adaptations for survival. Animals are self-determining in a collection of communities or ecological niches and habitats because they know what they need above simple instincts. Resources are fairly divided in separate areas, so no one is abandoned. Forest inhabitants, while self-determining, don't often make poor judgments or mistakes about their health, where to go, what to eat, with whom to associate. Those errors are predominantly a human problem via liberal policies of choice. In the vast forests there are loose-knit and varied species communities interacting on an ecological level. Self-determination is crucial in a forest since each animal, after some parental training, knows what's best for survival and reproduction. Governance per se comes from self-regulating Gaia that supports self-determination in species communities. Kymlicka (2002, p. 216) says, opposing the extremes of Mill (individuality) and Marx (shared experience) that what we deem as good is neither completely unique nor universal. Forest sovereignty for planet sustainability equates to a shared human and animal ethic.

For communitarians, the self is focused in society. While there can be social identity among many species, like apes and elephants, this might not necessarily be so in other animals and proves how no single human political theory applies to the forest. Yet the state of Gaia is a high-functioning political economy. Freedom of choice in the forest is both liberal and yet communitarian. Animals hold lives of value since their commitments are for relationships with others and especially the forest. Some species might be semi-solitary, as are some human religious groups, but unavoidably encounter others and nevertheless contribute to the forest ecosystem. Kymlicka (2002) discusses the notion of self in terms of other thinkers, for example the liberal freedom of Rawls v. the social construction of Harvard philosopher Michael Sandel; but there's no biology or talk of evolution or the adapted mind in these words. Species groups function tolerantly among other species, each recognizing the other's agency, what Kymlicka (2002, p. 231) calls the liberal tolerance for group autonomy.

Species as groups have adjusted their resource needs according to the demands of others, so that ecological niches represent, analogously, a type of utilitarianism of communities. Gains are shared for maximum fulfillment, and so forests could be left to reach their full potential without any or with very little human management. Kymlicka (2002, pp. 245–46) claims that any political philosophy must address questions of how an individual develops her autonomy and social condition under which she is able to be autonomous. Against liberalism, communitarians are concerned about the active and not the neutral function of these two points in a community culture of social good. While forest creatures have been free to adapt physically, mentally, and culturally, at the same time the state of Gaia has enabled them to seek some exactness of their cognition, intelligence, and social behaviors for survival and reproduction. There's a balance between extremes of liberal independence and communitarian identification with a group. There's participation in a community without loss of self-determination. The cultural communities of species might differ, but all have a sense of what's right for the forest in daily affairs for "common good," as Kymlicka (2002, p. 253) notes, recalling Peter Kropotkin from Chapter 1.

The forest state of Gaia is large enough, though shrinking, to support a distributive justice across animal communities in an evolutionary way humans cannot understand. While we were and still are prey for large animals, we consider ourselves hunters. Feeding is part of communal interaction in the forest. While individuals matter and have agency, what also matters is the net result of ecological engineering by forest flora, fauna, and fungi for climate health locally and globally. It seems that despite the niche divisions in Gaia, all organisms participate politically. These citizens accept the borders of their state that existed originally and have been minimized by human infiltration. On some level, forest animals know they belong to the state of Gaia, their identity. Only when forests became reduced did some, like primates, bears, deer, wolves and small mammals wander out of their diminished territory in curiosity and confusion. It's fair to say there's a forest identity for those in Gaia despite their multispecies cultures and behaviors. Their labors benefit many nations. Because resources have become impoverished in forests, justice dictates that they be restored through conservation and preservation.

Wolff (2006) assumes that because of human social nature we are, over individualism, communitarians. True enough, but the same applies to primates and many other forest animals from some species of insects to wolves and elephants. They are all social individuals. In Darwinian thinking, the individual organism would have preceded a collective society but for the mother-infant dyad in most vertebrate species. Curiously, political philosophers seem to ignore these facts and see human society *de novo*, assuming we are not naturally free nor connected to a long hominin and even mammalian past. That stilted mentality is what accounts for the presumptuous control we place on the forest state. There are moral implications between human societies and forests. People formulate and adhere to laws of the human state, so customs and traditions exist. Now is the time to use these rational and somewhat abstract laws to benefit the reality of the forests and all of their inhabitants. If justice is about fair treatment, and we don't have it fully in most societies, it's no surprise that we don't see it in how we assign our rights over forest animals.

Morality, Law, and the Environment

Human interests in land or water can be violated, but generally, trees and wildlife do not have legal rights of their own, Christopher Stone (2010) reminds us. If there's a lawsuit on behalf of trees, the proceedings would have to come through a guardian or trustee, Stone says. We have corporate personhood but nothing similar for the working inhabitants, nor for the entire domain, of forest Gaia. That's unfair since corporations can legally destroy vast swaths of living organisms and their habitats without recourse from nature for those damages. Stone outlines as follows the ingredients for a holder of legal rights. 1. The entity can file suit. 2. The entity can seek court remedy where injury has occurred. 3. Legal relief must benefit the rights holder. Stone says that environmental entities do not hold these rights, but they can be granted to a guardian or trustee. The holder on behalf of nature must come from a person who shows a violation of rights. As for number 1, rather than a vague individual suggested by Stone, let's recommend a group of ethical philosophers, environmental lawyers, ecologists, Indigenous people, and community leaders. As for number 2, the rights of forest flora, fauna, and fungi are diminished under those of people or corporations because they supposedly do not serve a public interest. This attitude is wrong, noticeable from the evidence presented in this book. As for number 3, positive feedback synergies in nature establish its self-regulation and maintenance, so when disrupted there is loss when there would be benefit.

Deforestation destroys unity, so preservation or reclamation with replanting would attempt to reestablish wholeness. In the current system, any monetary damages recovered do not go directly to forests but to plaintiffs. For Stone (2010, p. 6), there's a question of who legally speaks for wildlife, and he suggests a court-appointed guardian. The assumption underlying guardianship is incompetency. As outlined for the representative scientists covered in Chapter 1, the forest state of Gaia is quite capable and efficient. Courts are not much help since they are, in many countries, grappling with the question of personhood for animals using antiquated legal principles and not science or even common sense. Stone agrees that wildlife can indeed speak about needs and wants, as Eva Meijer (2019) has also discussed. Look, for instance, at the drought-stricken plains or the flood-ravaged mountainside to hear the cries about global warming. There is direct and not indirect loss to the environment. Even so, these problems are not best suited for lawyers or judges alone but for a cabinet of ethicists and biologists who study these issues. This assembly, including Indigenous people, might hold legal and decision-making power over corporations and politicians. The cabinet would enforce the notion of forest sovereignty through trust money and satellite imagery that monitors perpetrators.

As discussed in the previous chapters, we have placed monetary value on property rights, but there's much more at stake. Or, as Stone suggests (2010, p. 14), there could be a tort claim against anyone who violates the property rights and interests of wildlife. An easier and more beneficial approach is to set aside vast tracts of forest to be left alone, and only then would there be the rare case of forest property rights. That might be a naïve idea considering how nation-states have polluted internationals waters, the upper atmosphere, and even deep space. From Stone's legalistic perspective he's assuming there's damage; from another, let's avoid that from the start. In a lawsuit, lawyers and economists might speak of lost earning power for someone harmed. With damage to forests, the earnings get translated into two factors: loss of ecosystems to replenish and maintain themselves; forfeiture of vital air and water purification to all life forms via forests. Still, damage estimates are important calculations in helping people grasp the enormity of the problem and can be measured and litigated through the U.S. National Environmental Policy Act, which has a provision recognizing worldwide environmental concerns.

One might legitimately wonder whom legislators represent. While the answer seems to be their constituents, those people also have rights to clean air and water, whereby the forests have rights to cultivate a healthy environment for themselves and others, as even Stone implies (2010, p. 21). It's time for humans to use language that explicitly enunciates the rights of forest flora, fauna, and fungi. Some will agree to and others protest this and similar conclusions drawn so far. Either way, profound changes in how we view the environment are in order, which will impact all aspects of social life, from the individual up through the institutional level in consciousness and conscience. We can continue to pollute so as to reap economic gains, banking funds for our children and grandchildren; or we could protect the environment now, at some cost, for the sake of our children and grandchildren. Understandably, there are many developing and economically struggling nations that chafe at environmental concerns when their pressing issues are about boosting their economies. At the 1992 Rio Earth Summit, as Stone (2010, p. 116) reiterates an earlier point, the South resisted pleas from the North to deforest less. On the one hand, short-term thinking seeks to improve economies; on the other hand, long-term thinking of scientists is to stave off the worst effects of the climate crisis by, among other actions, reforestation and preservation.

As the climate crisis worsens, those with extraordinary wealth will survive better than those without, so the short-term thinking might not be the best approach. Instead, rich countries, as they have offered, can help developing nations since worldwide warming involves social justice. This means we need to enforce more than ever the notion of a global commons for all and climate as a social good. The environment is shared, and the pollution of one country damages the air, water, and soil of another. Deforestation is not a national but an international concern with the loss of biodiverse systems that act as gigantic carbon captures as well as air filters and water hydrators. For the poor nations, as sometimes occurs, huge forest tracts can be purchased by rich nations as sovereign states and overseen by a cabinet of philosophers, lawyers, scientists, and local people to ensure their longevity. Debt from poor countries could be forgiven in exchange for sovereignty of their forests. We need to move beyond U.N. declarations that sometimes carry little weight and are unenforceable to real solutions to preserve and expand forest areas.

U.S. Senators Martin Heinrich and Thom Tillis (2023) introduced legislation to expand forest and wildlife protections, laudable, but also deplorable that we have to legislate an ethical given: conservation. Questioning how little there is left to preserve, some researchers (Watson, et al. 2018) point out that over 77 percent of land mass, not counting Antarctica, has been altered by insalubrious human activity like cattle ranching, urban or suburban development, mining, logging, etc. Without quite saying so, these authors suggest that human overpopulation is pushing into disappearance and extinction massive amounts of wild species from flora to fauna. Like others, they confirm that some nation-states have legislated ostensibly protected areas, but these parks and forests are often encroached upon, at times with the permission of elected officials.

Animal Rights, Zoopolis, and Biopolitics

In their controversial book *Zoopolis: A Political Theory of Animal Rights* (2011), Sue Donaldson and Will Kymlicka say an ecological approach to animal rights, along with a focus on animal welfare, has failed. This is because ecology is viewed as an abstract whole without the individual animals and prioritizes human "ecosystems" as in deer hunting or wolf culling. Donaldson and Kymlicka advocate for animal rights but complain that it consists mostly of academics and is not politically engaged. In agreement are Alice Crary and Lori Gruen (2022) and Martha Nussbaum (2022) who argue for tying animal rights to social issues with more people taking collective action. Animal rights theory, Donaldson and Kymlicka believe, should weigh less on the negative (e.g., animals should not be seen as human food) and more on the positive, as in promoting forest animals as ecosystem engineers fostering global climate health (Tague 2020). Donaldson and Kymlicka (2011, p. 8) say it's misguided to separate the practical and relational obligations humans have to animals.

Justified or not, zoopolis has its detractors. Dinesh Wadiwel (2015) is critical of zoopolis because he sees it as too liberal and akin to Rawlsian theory, which as noted in previous chapters has limitations about animal life and human-animal interactions. In some ways, says Wadiwel, zoopolis also turns animals into commodities as bodies bent to human will. While that statement is true, Donaldson and Kymlicka and their critics don't really deal with the totality of earth's life as Gaia. Wadiwel's position is that violence and domination are inherent in the sovereignty of human politics, institutions, technology, commerce, etc. Drawing from Locke in his Chapter 4, Wadiwel says humans don't just engage in land acquisition but force their willpower over nature. As echoed in previous chapters here, biopolitics, Wadiwel believes, must not consider nonhuman forms of life as disposable property. He asks that we present the status quo with counter conduct by offering peace to others branded as animals whom we have colonized by abrogating their geopolitical power. Wadiwel says we've created our own self-caring ethics only after claiming sovereignty over others; we grant welfare after dominion since that's convenient for us (2015, p. 22). He goes on to say how globalization has conveniently situated power brokers in developing countries colonizing nature's domain. Where Wadiwel asks us to imagine the life of Indigenous people so that we can honorably negotiate with them, I say the same for forest flora and fauna.

In ways similar to Wadiwel, Paul Nadasy (2016) and Erika Cudworth (2025) are critical of zoopolis. Nadasy thinks zoopolis is biased against Northern Indigenous people and how they interact with the natural world. Indigenous people do not need a zoopolis concept. First Nation status imposed on Indigenous people is a form of colonial political imposition and often ignores their relationships with nature and wildlife, says Nadasy. Zoopolis argues for animal citizenship in a universal manner that makes the idea incompatible with the natural world. Cudworth goes further by saying that zoopolis is a form of animal colonization. She prefers practical applications that do not hedge species into compartments. For example, pandemic discourse against animals is capitalistic and a result of colonialism that recalibrates physical borders for humans and animals. Like Wadiwel, Cudworth employs Foucault's biopolitics to empower animals. At the same time, in modern human culture a pandemic fosters zoonotic fears and regulations that limit nonhuman lives. She argues for what she calls creaturely politics recognizing the rights of animals and their proximity to us. She would open boundaries, not restrict human-animal relations to reinforce Western colonialist inequalities. This thought does not exactly address flora and fauna deep in forest cores as argued here. Bernd Ladwig (2015) is also critical of zoopolis with its focus, using human political norms, on individual animals and the lack of principles regarding wildlife.

Like these critics, my argument is not for or against zoopolis; I go further asking for forest Gaia, which includes but is not limited to animal rights, sovereignty, suffering, etc., on a much wider territory (Milburn 2020). Michel

Foucault (1991, pp. 258–59) notes that the privilege of a threatened sovereign in ages past was the power of deciding someone's right to live or die. He goes on to say how with the disappearance of kingly rule and a rise of national administrations there's now pressure exerted on the "social body." Whereas in modern, administrative states the subjugation of children, women, minorities, and the poor can be conniving, the absolute authority over the life processes of nature has been blatant, what Foucault would call bio-power or when biology is subjected to political control.

Let me elaborate and situate myself more particularly, as if I have not done so already, drawing from Donaldson and Kymlicka (2013) in their defense of animal sovereignty. Along with inherent rights, wild animals have dominion over their communities and territories; by extension, I say this sovereignty also applies to flora. Wild plants and animals have earned and inherited by adaptation through natural selection membership in their forest ecosystems, a right not to be violated. So, the difference for me regarding the others cited here is that I include flora, fungi, trees, moss, lichens, etc., as rights holders to their sovereign, communal ecosystems evidenced by the evolution of Gaia's natural systems that happen to include microbes, insects, birds, reptiles, mammals, etc. Forest sovereignty is not about pets and less about individual species and more about connected biomes in an evolved biosphere. Humans are members of Gaia, too, and have a duty to be fair and just to wild plants and animals as a collective. Donaldson and Kymlicka (2013) imply that politically (where I might say evolutionarily) flora and fauna are rooted in and bounded by their communities and territories from birth, a form of sovereignty over human incursion. Evolved feeding behaviors, what humans call antagonistic predation, is simply part of multiple functioning systems of Gaia. As Donaldson and Kymlicka (2013) ask, what is the moral purpose of sovereignty. The answer, often repeated here, is to permit Gaia's abundant earth to flourish in its climate cleansing and water purifying functions. Besides, it's silly to talk of predation when virtually all humans at some point partake of animal agriculture. By virtue of its long existence, nature can care for itself without human intervention.

Kymlicka and Donaldson (2014) say that humans have different "moral obligations" for certain animals, whether domesticated or wild. Once we create a relationship with a class of animals, like a dog or a cat, we've established a fact that becomes moral. Relations need not be proximate to establish a moral duty since we should more than simply care about animals, plants, and microorganisms that service the climate biome we call Gaia. Animals live among us in our cities and towns as birds, bees, bats, rats, mice, squirrels, geese, opossums, feral cats or dogs, deer, coyotes, rabbits, etc. We interact with them implicitly and explicitly. That's the focus of these authors and how animal rights theory tends to separate humans from the reality of animal interaction. Wildlife far from an area inhabited by humans is a slightly different case. Moving away from abstractions and grounded in practicalities, we have a duty of care for forests to treat them with justice and fairness. Kymlicka and Donaldson (2014) agree to nonintervention for wildlife, but not for those animals in or on the edges of the human sphere.

There's a legitimate question about physical boundaries. Jane Goodall (2001) says that the area of her chimpanzee research at Gombe Stream National Park, Tanzania, has been reduced to a "tiny" 30 square mile plot compounded by human incursion where chimps are "isolated." Local people are farming on lakefronts, and top soil is washing away because trees are felled so others can farm on slopes, the very edges of the forest. The local people don't see this activity as destructive; they are only trying to survive. There are forms of remediation that would be better for the local people and the forest inhabitants, like fruit and lumber planting across many villages. Certain trees cut down indiscriminately for charcoal can be grown elsewhere rather than taken from the forest. There is much hunting in the area for meat that goes beyond any subsistence level. That could be corrected by small game farming or better yet, for the sake of animal lives, growing native protein plants. Deforestation and farming by local people in many countries on the margins of forests pose a real threat when multiplied. On scale, compounded with corporate deforestation, the situation is even worse.

Human and Animal Proximities

With increasing human populations near already minimized forests, wildlife crime is increasing (Mayer 2021). We should promote mental and physical health in relation to nature, not the opposite. Indigenous perspectives, on the other hand and from which we could learn, include: relating to and not separating from nature; spiritual beliefs grounded in the natural world; living sustainably; stewardship to the earth; holistic approaches to wildlife and ecosystems (Booth 2003). Human developments reduce animal ranges and movement, affect feeding behaviors, and upset population scales (Whittington, et al. 2022). Thomas Butynski (2001) says that any business operating in tropical forests is obligated to exercise an ethics regarding other species, especially those on an endangered list. He is not the sole voice blaming human population expansion for threats to wildlife and their habitats. For example, see Herman Rijksen (2001) and Colin Allen (2001). The world has seen how sanctions against rogue or aggressive countries can effectively limit their hazardous reach, so we might want to apply financial, media, or public pressure to help halt climate change, no matter how large the offending nation-state.

Inspired by ideas from Butynski underlining the thrust of my overall project, recall how in my Preface this proposal was offered. Local activists and young people could form a citizen panel, parliament, or community trust including Indigenous persons, guided and informed by legal, science, and ethics experts to inform their community about the urgency of forest protection; then, the citizen stakeholders could influence local, regional, state, and national elected officials to take action initiating legal forms of forest sovereignty by means of conservation, rewilding, or preservation. Consider this overarching proposal with the following specific, practical suggestions. Bradshaw (2020) is helpful here since she imagines, to be considered with some of the following ideas, animal or other forms of trusts, conservation easements, institutional investments, monitors, partial ownership rights to wildlife, and giving pledges or donations by extremely wealthy people. Tax incentives could be offered to private property owners who preserve, conserve and rewild lands. With a new cultural shift, we can only hope that credibility is bestowed on government land policies advantageous to forest sustainability. Some of these ideas are not original, but they are listed to remind people about what we could accomplish and as a stimulus for further thinking.

1. On a local level, sustainably use natural resources through a public, community declaration possibly leading to a legal decree. In turn, depending on the effectiveness of this suggestion, try to enforce the pronouncement as a national writ. Some countries have political "green" parties, but their platform of ideas could use continual reinforcement and expansion with reliable in-person and media outreach.
2. Local leaders, governmental finance branches (e.g., small business administrations), local banks, etc., might help prevent the financing of groups who violate number 1. Violators who willfully and illegally deforest or engage in unethical practices toward animal habitats and ecosystems could be penalized and defunded. On a global scale, the World Bank, the International Monetary Fund, European Union, etc., could be more involved with existing U.N. bodies. With the help of local law enforcement agencies using modern technology, like satellite photography and drone images, legal action can be taken. A public hotline could be established to address questions and help identify biodiversity and wildlife offenders.
3. Scrutinize any form of land development by commissioning non-partisan experts, partnered with a community panel with no conflicts of interest, in advance of groundbreaking. Typically, this now occurs through public hearings, but more outreach through various media could be done well before any meeting to sensitize the community to issues regarding

deforestation, water table disruptions, stress on infrastructure like sewers, roadway construction, potential runoff or flooding from superstorms, etc.

4. With financial incentives from private sources, philanthropists, wealthy countries, etc., localities and nations could expand forest remnants and national parks. In the U.S. alone, millions of dollars are funneled to election campaigns; imagine how some of that money or similar funds could be used for conservation and preservation efforts to mitigate the climate crisis. Political operatives tapped by a community panel could be employed using their fundraising skills and networks to help in this particular area.
5. Prevent logging and sport hunting, especially in newly created forest border zones; the community panel working with neighboring panels can pressure politicians who work with land developers, corporate AG, mining companies, etc., to be more conservatively minded. Indigenous people must participate in this effort and be able to retain their hunting and fishing rights.
6. On local, national, and international lands, confiscate with financial penalty any illegally hunted animal meats from a forest, savanna, or protected area. These laws exist but are not enforced; public voices working with community leaders and legislators can pressure law enforcement to act.
7. Alternative food resources for local people can be developed from and transported into areas using technological ingenuity. This can work on a local level, for example, urban farming, or internationally if funded by wealthy nations. Currently, billionaires are financing space exploration; some of their money, with proper financial or political incentives, could be used in establishing a veggie food enterprise.
8. Environmental education, conservation, and sustainable farming can become part of school curricula and community development, financed by rich countries who have profited from forest colonialism. Sustainability education should be obvious to any political party and can start on the local level in grade school even in the form of hands-on workshops. Funding for this low-cost initiative can come from any number of environmentally minded sources and even from local and national education departments. Legislators, with public voices, can be pressured to create such funds, which no doubt would be tiny compared to some police or country military budgets.
9. Biodiverse programs rather than logging and hunting could be developed and funded by corporations working in forest areas. Examples of this can be gleaned from a giant chemical company, BASF (2023), which is helping to promote the farming of raw materials in renewable cycles. Some researchers (Augustine, et al. 2022) have accused BASF of greenwashing,

a real concern with any corporation. Transparency of corporate activities and intentions regarding forests, agriculture, mining, logging, etc., are important. Young people with careers in management or finance could be involved with this effort, reading corporate annual reports, attending stockholder meetings, and asking questions of CEOs.

10. More products could be certified as sustainable and renewable; if not, countries could refuse to purchase them. This, too, is a challenge for greater transparency; corporations could benefit by increasing the public image of truly going "green." On the other side, conservation lobbyists could encourage negative news about corporate practices that harm biodiversity to arouse public interest in preservation.
11. Support of reforestation, conservation, prevention of destruction to wildlife habitats and animals, etc., could come not just from nations and corporations but from all sectors including local governments and the general public in the form of awareness campaigns since every living entity is a stakeholder in the fight for climate stability. Public consciousness could be informed by reliable sources, not corporate or political propaganda that often finances print and visual media promoting lies about deforestation and climate change. National and even local news sources could strive to make the climate crisis a lead story since it usually is not. More college campuses could hold environmental justice meetings, Earth Day events, vegan eat-ins, farm animal sanctuary information sessions, etc.
12. Flowing from numbers 10 and 11, big consumers of timbers, like Asia and Western countries, could be sensitized to forest depletion and biodiversity loss. Getting college-age people or recent graduates involved in public information efforts could be beneficial. Many college professors across a range of disciplines already are involved but need a larger platform to reach legislators and developers who will hear and possibly share their concerns. To wit, helping their own public image politicians could host events featuring teachers, professors, and students who engage in fighting climate change, biodiversity loss, and deforestation.
13. Bolster the efforts of organizations like the Ecology Trust (2023) to establish enforcement, monitoring, and remediation with money going only to truly sustainable businesses in countries making a concerted effort toward conservation. Some individual nations working toward this goal include Jersey and Panama.
14. Ethically attune and regulate ecotourism to assure flora and fauna safety while simultaneously not penetrating deeply into any forest area. Ecotourism money could flow back to creating sustainable farming practices on the forest edges for local people. Address disadvantages of ecotourism, like

stress on animals, trampling of habitats, disease transmission, pilfering of funds, etc. (Wrangham 2001). Ecotourism need not cease, but there might be ways of perfecting the practice with public awareness and the sanctity of wildlife in mind.

15. In developed areas prone to consistent flooding by climate change, consider state buyouts to relocate human populations and let the areas naturally rewild.
16. Private property is not an evil; offer incentives for large forest, wetland, and grassland owners to keep their lands green with biodiversity; preserve state lands at all costs and keep them free from development or mining through legislation; with tax incentives encourage corporations or billionaires to purchase large tracts of forest and hold them as wild in perpetuity. One such venture roughly related to number 16 is the Orangutan Land Trust in Borneo.
17. As alluded to previously with numbers 8 and 9, nation-states could use their military personnel and budgets not to deforest and pollute lands but to become active participants in conservation and preservation. Existing legislative committees could play a role in this effort with input from community voices.
18. There are many worthy environmental and animal causes and organizations to which one could contribute or volunteer expertise. Here are a few additional brainstorming ideas. Start a conservation blog; use the literary, visual, and vocal arts to support forest sovereignty; encourage book groups to read works that foster ecological ethics; talk to your children; encourage teachers to plan a class trip to a farm animal sanctuary; inspire school cafeterias to rely much less on meat and dairy; write to your elected officials; attend and speak out at community meetings; lobby homeowner associations to become more attuned to concerns of plants, insects, and animals; support and help promote lobbyists and environmental groups that promote biodiversity; form a group of friends who share a passion for wildlife preservation and tackle some of the suggestions noted above; while listening to others be prepared to respond. One can live an environmentally ethical life in many ways to influence the moral thinking of others.

One takeaway from these points is that corrupt commercialization of forests is not real conservation, to the detriment of earth's health (Ammann, 2001; Wilkie, 2001). Many people tend to view wild animals as caricatures separate from humans and not as subjects of their own lives in forest habitats. This attitude can be remedied, say Tara Stoinski, et al. (2001) through education, especially repeated visits to zoos. As they note, there could be habitat conservation lessons before a zoo visit followed by persistent education focused on animal rights and the importance of forest biodiversity. Many school children, indeed their parents, see zoos as a form of entertainment, not significant instruction about the importance of animals in fighting climate change. Roger Fouts (2001) says there's a lot of misguided biology and twisted ethics concerning humans and animals, a dichotomy which does not really exist but has been perpetuated in education up through college by misreading philosophers.

These reflections were necessary to expand on and conclude a close reading of the political philosophy from Chapters 2 and 3. Next, let's consider how philosophical ideas and cultural evolution hinge on "humanity" in this debate. A thought to hold is whether or not humanity collides with nature. Contrary to popular belief, we are not an exceptional species, and by now readers see how much damage we've done to the planet. Exceptionalism and speciesism are difficult ideologies to break. There's also too much emphasis on technology in our culture and how it might save us.

In terms of culturally inclined thinkers discussed next, Hannah Arendt's conception of human nature aligns with a picture of self-interest. Kumer and Campbell's evolution of morality poses a similar problem about misconceptions of human success. Again, we see how anthropocentrism leaves little room for deliverance through ecological ethics. Likewise, Martha Nussbaum advocates for animals but reveals the limitations of narrowly conceived notions about their capabilities. These authors pave the way for a Conclusion that addresses questions around the politics of Gaia in an atmosphere of socially learned ethical ideas.

The Lure of Science

In 1958 at the dawn of the space age and about ten years after the nuclear age had begun, philosopher Hannah Arendt (1998) warned how passivity could be tempted by science, taking us away from contemplating our actions and goals. Arendt bolsters an ecological critique of Enlightenment and social contract theory. We face a similar temptation today with ideas about technological solutions to climate change. After World War II, Horkheimer and Adorno (2002, p. 31) also warned of privileging an object over the subject. Like Marx, Arendt sees how individuals have been disillusioned in a society dominated by economic and technological machinery. For example, these days artificial intelligence can generate and spread misinformation about climate change to unaware or passive readers (Marcus 2022). Consumption and industrial production along with overpopulation have moved people away from a communal, participatory democracy, blinding them from taking responsibility for the consequences of their actions. We have the power to alter damaging attitudes through cultural evolution. Arendt's book is a study in how some aspects of being human are stable (adherence to tradition) and how some are flexible (reception to change).

Humans have created an artificial culture separate from other organisms, seeing our animal bodies and minds as imprisonment, Arendt proposes (1998, p. 3). She goes on to say that humans by the late 1950s developed the ability to destroy all earth's life. Arendt suggests that what distinguishes humans from animals is our ability to work socially (1998, p. 22). That's not totally accurate, for nonhuman forms of life do not simply labor in solitude but contribute their efforts to Gaia that contains and connects all organisms. Besides, there are many forms of social activity in nature, evident in wolves and elephants to ants and primates. As demonstrated in Chapter 1, some researchers like Simard and Wohlleben might say trees and their root fungi are social. Arendt notes that in the Middle Ages the common good was not political but indicated an individual's interests held in common with others (1998, p. 35). Similarly, the border for forest organisms between home and the common sphere is blurry.

Arendt seems more about moving human behavior to public excellence, but her public realm is privately human and not in nature. The human public space is not one with nature but erases wildlife. She makes a distinction between labor, or what is performed biologically, and work, or the political essence of collective individuals. Labor is like slavery to administer life's necessities while work is about completing a product (Arendt 1998, p. 80). Like the other political philosophers outlined in this book, with some exceptions (e.g., Sue Donaldson and Will Kymlicka), Arendt assumes that animals only labor in acts of consumption while humans work by using tools and the arts. This is untrue, for the work of forest ecological engineers is a social good on the authority of Walzer (1983). Many creatures use tools, from corvids to primates, and great apes find pleasure in painting. Carol Gigliotti (2022) wrote a book about the creativity of animals. Work is not, as Arendt imagines, inimical to the natural world; though for humans, work drains matter from nature without any remedy. Contrary to thermodynamic equilibrium noted above in Chapter 1, William Rees (2009) sees technological humanity as a "dissipative" system degrading the material energy in nature.

Locke, Smith, and Marx make no real distinction, as does Arendt, between work and labor. They see labor as a source of property or value (Locke), wealth or exchange (Smith), and productivity or physicality (Marx). Arendt admits, though, that cycles of labor, rest, and consumption are natural and favor an individual's needs. From the human perspective, forest workers are indistinguishable laborers (her usage), but their actions across species are really both individualistic and cooperative. This point is evident with the human elimination of niches, habitats, and ecosystems creating a negative cascade. Individual organisms in nature effect a collective, cumulative energy of inestimable fertility in forests. If forest organisms are workers and not just laborers (by Arendt's definition), then they are free and deserve consideration of justice. In contrast to the balance of work and labor in the forest, virtually all human societies are engaged in a "waste economy" (Arendt 1998, p. 134) of endless consumption, energy loss, and garbage production. This culture of viral wastefulness is very different from the recycling of matter by forest organisms.

Work is more to hands and fabricating something of property (Locke) or value (Smith) compared with labor using the body (Marx). Misconstruing the obvious, Arendt (1998, p. 139) says animals labor as servants to the earth while humans have mastered nature. This view ignores how nature precedes and will outlive humans. No matter what we do and if we destroy planet earth, especially if we try to engineer the climate with technology not considering the consequences, some microbes will persist and begin anew the processes of evolution. Arendt admits we have de-natured the world with forest extinction and have created with tools our artificial world separated from and destructive to nature (1998, p. 148). The problem with work for humans is that we think in terms of means and ends where everything has tool potential for transactional utility, and Arendt seems not to advocate this approach. In modern human eyes nature has no value unless it becomes an instrument of use. Most societies judge people on the "value" they place on their work in an exchange market. We should therefore assign importance to forest workers who sustain and regulate Gaia. Arendt is not a utilitarian and has a sense of how modern people enslaved by technology have abused and devalued nature without making a better world, or, more precisely, have put a monetary value on ecosystems.

In human terms, Arendt (1998, p. 198) notes how the *polis* is conceived of as people living and working together, not necessarily defined by geography. Forest workers live and act together, are not defined by specific boundaries but by the holism of Gaia, and thus deserve sovereignty. Forests should only be colonized by plant and wildlife, not humans, because of their prehistoric and productive animation in these spaces. Not surprisingly, contemporary humans place high income earners above many who perform physical labor or are engaged in the arts or teaching. This erroneous perception places forest inhabitants, who maintain Gaia, even lower on a false hierarchy. In reality, forest wildlife engages in rhythmic cycles of exchange where nothing is wasted or where one has more than another. There's no alienation since complex relationships are at the root of Gaia's natural processes.

Arendt (1998, p. 245) suggests that sovereignty is the purposeful action of many working together, a main point concerning forest workers whose singular efforts collectively keep ecosystems alive and well for the masses. Sovereignty can be gained, lost, and limited as we see in human realms, but wildlife has earned that right in their drive to preserve the present so as to ensure the future, proved through the continuity and evolution of species. The moral dwelling of humans is no longer the forest since our increasingly bipedal hominin australopithecine ancestors left the trees millions of years ago. The human condition is a drive to advance with politics and government, industry and finance, science and technology thereby creating a mortal place from which there's no return. Arendt (1998, p. 250) calls this human condition the "shrinkage of space." Simply, we've stolen and abolished land, waters, and airspace from wild flora and fauna. As of this writing, there are about 195 countries whose citizenship is bound by space; these lines of demarcation don't but could apply to wildlife citizens who live everywhere. Wildlife is earthbound. Even so, birds don't presume to conquer outer space as do humans. The human condition is full of willpower contrary to nature's evolving growth of sustainable productivity; humans have lost a point of reference wildlife has not. In some respects, the doubting of Descartes turned philosophy away from nature and focused only on an inner, human life (Arendt 1998, p. 298).

The Question of Human Success and Capabilities

In *A Better Ape*, Victor Kumar and Richmond Campbell (2022) write in the shadow of Arendt and debate the notion of moral progress. In an anthropocentric manner they pin the success of humanity on morality and seem to question the direction in which we are headed. First, like other writers (Henrich 2016), they place too much emphasis on what they see as a special ingredient in humans. We are not exceptional, evidenced by how much damage we've done to planet earth. Second, the title of their book privileges humans. We are no "better" than any other great ape. In fact, by the end of their book they openly ask if we are indeed better, undermining their entire thesis. We are not superior, seen by the extensive and continual destruction we've done to other humans, plants, and animals. In truth we are, compared to orangutans, gorillas, chimpanzees, and bonobos merely a different ape. Kumar and Campbell insist that our evolved morality has made us human. This is exaggerated since the wages of humanity include our climate crisis on top of all the other problems we create, from poverty and homelessness to war. Humanity has contributed nothing to nature. Human capitalist, colonial attitudes to nature as a resource pit serving those with power and money now jeopardize all life. John Piccolo, et al. (2022) affirm that nature has rights because of intrinsic values we must acknowledge according to "ecological justice." Ethics cannot be based on the notion of human exceptionalism; we are not morally above any other uniquely evolved species.

While biology and genes play a role in human social development, culture is the linchpin. Whether we are in Argentina or Zimbabwe, we are unavoidably surrounded by culture from parents, grandparents, peers, educators, and even adversaries, with extra force exerted on us from the arts, sports, media, etc. Kumar and Campbell (2022) seek a theory that touches all other notions of human evolution. Morality, they say, was a prime mover of human genetic and cultural evolution because it's tied to cooperation and intelligence. Culture became a means of self-domestication away from aggressive and cheating behaviors with the socialization of morality in groups for greater fitness, including widely accepted norms eventually manifesting in institutions. They break the center of morality down to emotions, norms, and reasoning, which they note makes morality pluralistic and not singular. Culture affects the outcomes of morality via family, peers, and groups. Morality does not necessarily mean becoming a slave to social cooperation for the greater good. Moral reasoning is merely second to one's reactive feeling in a situation but can play a role in framing social action. The upshot is that we don't seem to have made much moral progress. Following Arendt, we really need to think about our wants regarding our actions against nature. Human activity corroding the forests and the climate were well known in the final decades of the twentieth century, as previously noted, and little was done to adjust that culture. Now that we are in crisis mode, everyone seems alert. That's not moral progress but a fearful response.

Cultural evolution is still Darwinian with trait variation and the passage of differential fitness factors through non-genetic means (Tague 2022, ch. 5), like learning how to make and use tools or control fire. Imitation and social learning, apparent in other species, is nonetheless critical to the human cultural explosion. Especially for humans, cultural selection in social learning happens frequently when groups compete against each other, and less so through individual competition. Biology factors into cultural evolution since genes underlie physical and neural behaviors influenced by information as well as the environment. Individuals and groups who capitalized on fitness-enhancing cultural intelligence survived and reproduced better. Along with cultural evolution humans evolved norm behaviors of approval and disapproval, which are similar but can differ in some groups. The likelihood of social punishment helps most people adhere to moral norms, but prevailing codes enable pollution and disable conservationists. One study implies that student ethics are compromised by climate research professors funded by corporate agriculture (Morris & Jacquet 2024). The message here is for the creation and promotion of behavioral rules that are environmentally adaptive and beneficial, leading to further cultural evolution.

Expanded social cooperation through stabilizing norms helps cognitive capacities sympathetic to forests emerge, ratcheting up yet more cultural evolution in an environmental psychology. Bio-cultural evolution is important for adaptive social behavior and learning. Human norms that supported cooperation and sympathy and then solved disputes spread more widely among *Homo sapiens* since sensible rules were overall advantageous to everyone. Small groups of conservationists, whether academics, independent scientists, writers, musicians, or visual artists could cohere and help build moral emotions and ethical reasoning about forest conservation. In spite of better standards of living for many people, there are questions regarding the notion of moral evolution. For example, how inclusive are societies and how do forest flora, fauna, and fungi fit into any framework of moral progress? While we have seen progress in some social causes, Kumar and Campbell (2022, p. 183) admit there has also been moral "regress" in some areas, which surely includes ethical aspects of animal rights and environmental preservation. A subcategory of moral exclusivity that's persistent is moral inequality, best exemplified in the expanding corporate agriculture of meat in some countries (Ritchie, et al. 2017).

Borrowing from the ethical contemplation of Arendt and the moral flexibility of Kumar and Campbell, let's take this cultural evolution approach a step further. Martha Nussbaum offers a contemporary response that advocates liberalism for animals in *Justice for Animals* (2022). Using a capabilities approach, to be defined momentarily, Nussbaum makes a compelling case in her argument. Her strengths are in philosophy and the law, two pillars in a claim for animals to live and flourish (her word, loaned from Aristotle) as they so evolved. Nussbaum says our debt, overdue, is to heed voices calling for better treatment of animals, whether through welfare reforms or rights regarding animal agriculture. We are all implicated in various injustices to animals, from insects to whales, whether through habitat elimination with urban sprawl, eating meat from factory farms, pollutions that stain the environment, plastics that injure marine life, etc. Nussbaum believes that an ethical person would not harm any animal, like a fish, if scientists prove it feels pain. People also inaccurately underestimate the ability of animal minds while imposing a belief of human supremacy (Leach, et al. 2022). Because of direct or indirect human harms many animals do not flourish and suffer. Most animals are subjectively aware and so, conjuring Aristotle, it's unethical for humans to stifle how any animal needs and desires a full life. Nussbaum mostly faults law and political theory for their lack of addressing the problems of cruelty to animals, whether in captivity or the wild.

As has been argued, one strand of justice is political. Alasdair Cochrane, et al. (2016) offer a political focus of justice that is not geared toward morality since behavior might not concern how one should "regard" animals but how a state treats them. Or, in the case I've presented, grant wild flora, fauna, and fungi forest sovereignty. Human laws follow trends in values and beliefs, and so thinkers like Rawls, Nozick, and even Walzer are employed for something new, a theory of forest sovereignty. Representing the message of Chapters 2 and 3 here, Nussbaum believes that political theory is anthropocentric. Central for her is how animal justice is positioned in the ability to live fully and thrive as an individual or in a social group without harm or interference from humans. What she fails to mention is how animals, mostly mammals in her outlook, depend upon a habitat of plant life. Nussbaum asks if animals have been made able to flourish by human laws and institutions. Her answer is no; but changes in attitudes and laws can make animal lives richer. Provided they have enough fertile land, wild animals do flourish, as evolution has proved.

Though antithetical to Nussbaum's cause, the mission forward is to legislate rewilding or reforestation of newly created expanses of land; then, on the margins of those forests revised laws can help animals and humans. By her own account, Nussbaum says that animals in the wild require human intervention. This invasion is not necessary and what has depleted habitats in the first place. The opposite and recommended approach is to designate vast tracts of contiguous land as untouchable forests, which would include wetlands and grasslands. Justice for animals requires their ability to freely attain the potential of capabilities in a natural habitat. By default, this liberty will help the biosphere and, subsequently, humans, as a key component for stabilizing Gaia's climate health. In other words, we need more than restoration and the setting aside of old-growth forests. Future planning would include international justice and laws to avert the imminent "tipping cascade" of warming, arctic melting, flooding, drought, etc. (Kim 2021). This arrangement is not the same as paternalistic intervention for the sake of some animals from a clouded human perception of their needs. There seem to be no boundaries regarding injustice against animals, since human occupation of the planet is so pervasive, and at times perverse, that it overpowers wildlife. Animals, Nussbaum says, should be free to choose their "essential life-activities" (2022, p. 8) and so thrive in their environments. She fails to take this a few steps further. The word "animals" in her vocabulary seems to exclude plants, microorganisms, and insects vital to a thriving ecosystem. She favors vertebrates. She also fails to note how every living organism contributes to the whole of forest Gaia; there are no special inhabitants.

With cognitive dissonance many people who claim to love animals (e.g., their pet dogs) also consume large quantities of meat and dairy, two industries that wreak serious injustices to animals and the environment. In the U.S., a representative home emits per year 48 metric tons of carbon dioxide equivalents (Center for Sustainable Systems 2021). Meat and dairy account for almost 75 percent of that carbon footprint, and this total includes production and transportation. In

2019, in the U.S. alone, according to the Center for Sustainable Systems (2021), ruminant farm food like cattle, sheep, goats, etc., produced 179 million metric tons of carbon dioxide equivalents of enteric methane gas. There is no animal justice in corporate agriculture or human food culture. These are elements of the human condition Kumar and Campbell (2022) ignore but which Arendt (1998) predicted. GDP is good for measuring a nation's economic growth but might not reveal how individual lives are affected, positively or negatively. Thus, as an alternative to GDP, the capabilities approach was developed as a more equitable measure. Seemingly negotiating Nozick and Rawls, there are basic (innate) and internal (developed) capabilities which, under the best circumstances, combine for an optimal result as part of justice. For people, the key is to maximize education to minimize disadvantage; but even so, many animal lives are disadvantaged out of cruelty or lack of individual development, Nussbaum notes (2022, p. 87).

Here are the central capabilities, listed verbatim succinctly followed by my summary explanations. As a bonus to any capabilities approach, according to Walzer (1983), some of these qualify as social goods with different meanings and values for humans and "animals." Nussbaum does not make that distinction.

1. Life. One is entitled to a normal, healthy life span.
2. Bodily health. One should not be deprived of enough food or shelter.
3. Bodily integrity. Living freely and securely from unwanted attack.
4. Senses and thought. The liberty to be psychologically and mentally creative and vibrant.
5. Emotions. Independence to experience connections with others.
6. Practical reason. Understanding what's good in life and planning for goals.
7. Affiliations. Living without prejudice and with respect and dignity in a community.
8. Other species. Recognizing and avoiding harms to the wider sphere of nature.
9. Play. Time and opportunity for leisure and enjoyment.
10. Control. Engaging freely, especially in terms of habitat and property rights.

These capabilities were designed as threshold benchmarks of justice for nations to develop. Nussbaum (2022, p. 90) admits these are not as easy to measure as GDP. The question, away from Kant who said only humans have dignity, is how governments can legislate institutions to address the necessity of respect for animals given how Arendt (1998) sees them only as laborers and Kumar and Campbell (2022) see them as inferior to us. Justice for animals and their environments is a barometer of how humans interact socially. Nussbaum imagines a political consensus accepting the fundamentals of a capabilities approach with a few differences (2022, p. 94). Just as people deserve a chance to be the best they can as humans, so too with all animals who strive to fulfill the goals characteristic of their species. While she does talk about many species and their abilities, there is no discussion about them as ecosystem engineers. That's a serious omission. Humans in attaining capabilities are likely to erode environments and pollute. That's not the case with forest flora, fauna, and fungi. Nussbaum suggests that a list of animal capabilities can easily be established and possibly enforced by experts enshrined in a type of constitution. I assert somewhat differently the following. Because flora, fauna, fungi, microorganisms, etc., engage naturally in ecosystem engineering to sustain their biomes and by default the biosphere of Gaia, they should be granted forest sovereignty.

Nussbaum says the capabilities approach should apply to creatures with intentionality and subjective experience (2022, p. 119). This application would exclude many microorganisms, fungi, algae, lichens, mosses, etc., working in forests to sustain soil ecosystems. Nussbaum says a theory of justice need only apply to creatures with sentience or subjective awareness (2022, p. 138). She's skeptical but open to contradiction pending new evidence that plants are sentient (what she defines as intending or striving for) because there's no neuroanatomy (2022, p. 150). All plants move toward light to photosynthesize, a real response, as noted by Darwin in his groundbreaking books on plants and recently by Paco Calvo (2022). She does not see trees as individuals but as a "cluster" of seeds and leafing branches, so ethical attitudes toward them are more to a species and not as an individual. Such thinking ignores ancient trees like giant sequoias. While her focus has nobly been on individual suffering, ignoring the flora upon which mammals co-evolved is an error. Nussbaum bemoans how animals kill each other but fails to see how that's an evolved feeding behavior sustaining an ecosystem. A lion can't choose between an antelope and a cantaloupe in a supermarket. According to Nussbaum's capabilities approach death is harm to a conscious being who has a sense of place in the past and future (2022, p. 165). While this applies to animals, I'd go further saying that forests also are entitled to the capabilities approach since their thriving biome is directly connected to the flourishing of many individual lives.

Nussbaum is advocating for animals but is inflexibly against expanding and leaving alone wild, natural habitats. She justifies this position with repeated claims that there are few wilds left. She will not contend for larger natural areas, as she's already insisted that's impractical and will only harm animals. On the contrary, there should be a concerted international effort to expand contiguous forests and leave them alone as natural habitats. For humans, that means concentrating more people in sustainable cities and ceding back to nature land that can rewild. What Cal Flyn (2022, p. 26) calls the "latency of life" manages to spring up over time and establish living forms where once they were obliterated by humans, whether at a mining site or the location of a thermal-nuclear explosion. As Flyn cautiously notes, this is not quite restoration because of the massive destruction or pollution underlying the struggle for life. Rather, these sites exhibit colonies of adaptation. We don't need to curate or cultivate nature since such intrusion brings us back to the original problem. In North America, coastal temperate rainforests provide ecosystem nitrogen through canopy moss systems, validating the importance of old-growth trees in forest dynamics (Lindo & Whiteley 2011). All flora and fauna, as well as insects and microorganisms, constitute an ecosystem. How can we claim justice for only one part or a select few?

Conclusion

Ecological Credo

This book has addressed political and moral philosophy, animal ethics, and environmental conservation and preservation. In line with the ancient Greeks, and as I've attempted to elucidate, ethics and politics are entwined (Hadas 1950, p. 154). Much territory has been covered with the hope that readers will dispute so as to engage with the idea of forest sovereignty. Since philosophy is at the core of this wide-ranging argument, let's conclude with ruminations circling around and tying together some main points. Like Arendt after him, Bertrand Russell (1952) debates what is certain and questions if science has all the answers. Amidst uncertainty, philosophy can at least ask relevant questions. Physical science is useful in daily life, but as Russell indicates, anticipating Martha Nussbaum, philosophy adds value to one's mental life by tackling big problems and critically examining and unifying knowledge with inquiry and insight.

The Political Society of Nature

What matters is not only Nussbaum's focus on self-awareness of some species but more so how all parts, from the miniscule particle of soil under the ground surface to the canopy of a great tree, contribute to global well-being. We can't confine justice only to selected, sentient animals but have to shift gears and grant sovereignty to the life-sustaining and sapient forests of Gaia.

Ideas from political philosophers deal with the moral evaluation of societies. How does any society treat nature? In discussing "humanity," David Miller (2003) declares that people need laws and governments to cooperate, but we see cooperation, mutual aid, and care for others in nature. Nature does not need political authority to exist well. Gaia is a different type of state. Echoing others, the thrust of anthropocentric political philosophers is the focus on collective choices about values in people's lives excluding any consideration of wildlife. Some readers might say that my argument is consequentialist since I focus on actions as a function of utility. For wildlife, perhaps that is true; for humans, less so since their actions should be guided by virtue ethics in protecting the sovereignty of forests for the good of all life. Right action toward nature by essentially leaving forests, wetlands, and grasslands alone as earth workers creates a byproduct of utility. Usefulness is not the aim since that could justify wrong actions to achieve some ultimate and possibly unanticipated goal, as seen in how palm oil plantations were touted as good for the environment and yet are responsible for destroying Indonesian rainforest biodiversity.

A. John Simmons (2008, p. 7) defines society as peaceful and cooperative stability among inhabitants (he says "persons") whose varied conduct nevertheless adheres to some rules of right behavior. A problem political philosophers grapple with is the legitimate right to rule. Humans have taken for granted, perhaps from the Book of Genesis, that they have power over wildlife. Justice for humans is an advantage of the seemingly stronger over nature, to borrow from Plato's *Republic*. One can debate this form of justice, for one's strength can also be a weakness, as Marx tried to show in how the effete bourgeoise manipulates robust workers. Justice equates to equality, reward, and punishment in an older tradition, so circumstances and contexts must be considered in terms of what is needed and what is deserved (Miller 2003, p. 76). Human history is not filled to the brim with peace and cooperation, though we can learn about combined effort from our wildlife cousins.

In addressing the question about a transnational state of Gaia, one needs to consider the liberties and rights of forest inhabitants who own and control their resources without harming others, putting aside evolved eating behaviors since those are part of ecosystem cycles. In many ways, nature is nonpolitical, but humans have financially politicized forest plants (e.g., pharmaceutical companies) and wildlife (e.g., bushmeat trading and exotic species poaching). Simmons (2008, p. 17) notes that sovereignty is state property, so it stands to reason that independent Gaia, which sustainably maintains forests, wetlands, and grasslands, should be a state. Like others (e.g., Hobbes, to some degree Locke, and even Nussbaum), Simmons believes that nature is anarchy or the absence of government. Political authority includes a strong dose of compliance (Miller 2003), necessary because without it we'd have the nasty and brutish world Hobbes imagined. Yet, as a collection of very large and diverse bodies Gaia is self-governing in joint action. People are necessary for a human state as thinkers from Hobbes and Locke to Simmons seem to accept. Masses of people are not required for the natural state of Gaia populated by a range of free and self-sustaining plants and animals. Simmons, like Rawls and Nozick, asserts that the main benefit of a state is security. While some animals engage in adapted feeding behaviors against others, the truth is that myriads of forest organisms working together live untouched. Besides, feeding ensures control and stability of populations, and what's eaten cycles back into the body of Gaia's ecosystems. Social meanings and values determine the distribution of goods, says Walzer (1983), and we see that in Gaia. The invisible structures of nature without the prejudice of human thought are in a holistic balance, as they have been for millennia.

Many political philosophers like Simmons (2008, p. 20) investigate how human "social life" exists only in a sphere of people where rationality is preferred to the state of nature. Opportunities, assistance, resources, sociality, etc., also exist in nature across and between species. As for state security, many species have evolved adaptations from body armor and poisons to an ability for hiding or, in extreme cases, retaliating. There is no tyranny of Gaia over its inhabitants. A forest state of Gaia needs no human justification since it has improved the free lives of its workers by virtue of their evolutionary adaptations in an exchange with the ways of nature's environment. Much of the flawed anthropocentric thinking can be seen in how Nozick claims that competition in nature will result in one "dominant protective agency" becoming the state (Simmons 2008, p. 35). This attitude assumes systems in nature are unstable and that human structures are not, when the opposite is true, as even Simmons recognizes. There is no single dominant power in nature other than widespread energies, but humans often create states that consolidate power for a few and force others to follow, relent, or retreat.

There are political obligations in the broad sense of mutual obligations in a forest, for example avoiding physical confrontation or respecting another's boundaries. One might even say there are associative duties among forest inhabitants in, as Simmons (2008, p. 50) says, a "web of mutual obligation." Humans, too, have an associative duty to wild flora, fauna, and fungi given the strong benefits we've garnered in an unequal exchange. The problem is that we typically see associations as consequentially transactional. That's not really the case for forest organisms where the net result for all participants, like sustainability of soils, water, and air is what counts. There seems to be more state resistance among humans, rather than compliance by resident forest wildlife, considering the large number of global migrants. Simmons (2008, p. 62) avers that political philosophy must address both a state's legitimacy and justifiability, as shown for Gaia. Somewhat following Rawls, as explained by Simmons (2008, p. 80), there are multiple goods of many individuals spread fairly and not one "favored" good for a few. Forest workers on some level seem aware of everyone's free right to resources for the advantage of all, as Rawls might mandate in terms of justice. That sounds like a liberal society with stability originating from the many organisms and their differences filtering through the ecosystem.

If humans remove their anthropocentric blinders, clearly everything in nature fits with other pieces. As Simmons (2008, p. 90) points out, Nozick takes Locke's notion of property through one's labor as entitlement no matter the type of distribution. Compare humans attacking nature for resources with wildlife in nature maintaining and growing resources. There are striking differences in acquisition, transfer, and renewal. In nature, but not necessarily in human societies, there is acquisition with leaving enough for others, shared transfer by consensus, and replenishment of food and water. Moral complaints can be waged against humans, not Gaia. Injustice stems from how humans harm forest flora and fauna for their perceived entitled advantage. In nature there are market freedoms with fair exchange. This situation is not the case in human societies and especially not true of how humans treat nature. As Simmons (2008, p. 109) discusses Mill and Rousseau on voluntary association, tailored for my argument, individual wills and desires are satisfied in an environment that typically strives for the common good of all and not a privileged few. There's synchrony in nature, not usually seen in human societies between individual needs and the larger whole, for example trade imbalances and market instabilities. Attesting to the anthropocentrism of political philosophy, a thinker as sensible as Simmons (2008, p. 122) nonetheless says that in a democracy only people and not "lower animals" can be represented.

Simmons (2008, p. 128) says that there's no world government, but as strenuously argued here this is misleading since Gaia has ruled the globe for many millions of years. Rousseau might react by saying that what counts is willful sovereignty and not diluted representation (Simmons 2008, p. 123). Rawls would object to a common good since it might point to a state's agenda, but in the forest wilds the common interest is to support many lives, not a favored few. Whereas in a democracy there might be public deliberation of some regarding a common good, for wildlife there has been a process of adaptation involving all organisms large and small. For those who disbelieve Gaia's potent influence, consider the current climate crisis and its dangerous effects on the human economy and how hard it has hit all life forms. Contrary to much political philosophy, the state of nature is not in raging competition. Rather, humans have waged war with nature in their acquisitive competitiveness producing morally dire results. With political decisions, questions of fairness about how humans treat nature include moral implications.

Echoing Locke, Simmons (2008, p. 136) says any legitimate state has rights and control over "claimed" territory. This assessment does not include the rights or legitimacy of historical Gaia and its evolutionarily ancient inhabitants. Forest wildlife, to borrow terminology from political philosophy, has been an effective steward of its territory. Wildlife and plants in forests, wetlands, and grasslands are politically self-determining by virtue of an interlocking web of organisms forming a *polis* with connected habitats where individuals voluntarily strive to live in ecosystem communities. There are no lesser groups in the forest state, for the bee is as important as the plants it pollinates to feed mammals and others. Rather than nation-states we might want to consider, as Miller (2003, p. 112) says, political philosophy on a global scale with international justice. A similar idea was raised in Chapter 3 with Eckersley's notion of the transnational state. Miller acknowledges that environmental problems require an international response and should not be left to any particular nation. Though he focuses only on humans, a global ecological society would ensure rights to all forms of life.

A Natural Contract

In *The Natural Contract* (1995, p. 3), philosopher Michel Serres claims that human culture "abhors the world." There seems to be truth in how our destruction of nature is global and no longer local. Serres suggests that some have wagered to win a war over nature, yet any loss will not be theirs alone but forfeiture of many life forms. At the time of his writing around 1990, he says there was collective anxiety about the state of the natural world; since then, the situation has worsened. Darwin (2003) talks about the struggle for existence, where plants and animals scrounge for food and mates in a defined area. Blowing aside Darwin's theory of evolution by means of natural selection, industrial humans seek a teleological design of complexity in their machines, whether for drilling oil or coal or burning those fuels, with the result that people of industrial nations doom their own adaptation to climate change while forcing the crisis onto countless other people and nature's organisms. Modern humans exhibit a belligerence against nature where, in the words of Serres, we present a mostly unregulated "objective violence" (1995, p. 10). Many nations band together almost as disease to dominate nature. There is no social contract that strictly protects nature. There has been no agreement between modern humans and the transnational forest state of Gaia, no accord, though there are rays of hope from some countries like Côte d'Ivoire (biodiversity conservation with U.N. help in spite of wildlife trafficking and deforestation), Madagascar (reforestation and anti-poaching), Mozambique (protected areas and conservation laws), the Dominican Republic (public awareness and the Nature Conservancy), Nicaragua (reforestation and community conservation initiatives), Fiji (restoring coral reefs and marine conservation), and Lao PDR (working with the World Wildlife Fund to decrease animal trafficking and increase biodiversity awareness). Canada has much sustainable forest and holds leadership on that front in spite of wildfires and conflicts with Indigenous people (e.g., over land and resources).

Serres (1995) suggests that it's only human to occupy and overcome nature, which we see in large scale by the late twentieth century in massive deforestation. Years later, Serres goes further by asserting that humans claim an "unwritten right" (2011, p. 27) to act like parasites on nature, harming forests and eating animals. We have come to possess nature, he says, through a virus of "malfeasance" with our personal excretions, garbage, gas fumes, foul odors, industrial waste, etc. Serres defines hard pollution, or business waste, and soft pollution, or our collective power of will to possess lands through our literal and figurative filth (2011, pp. 41–2). In other words, we don't need science and data to see and measure this corruption; it's quite evident from our activities on personal and communal levels. Moreover, Serres asserts that possessive needs and dominating desires of wealthy people contribute the most to both hard and soft pollutions (2011, p. 45). Opposed to appropriators and polluters, and perhaps too optimistically, he says humans must see themselves as responsibly clean tenants of nature's world shared with other organisms, a natural contract of "dispossession" (2011, p. 73). Serres submits that organisms have a right to their natural habitats; ecosystems are not owned by any person, and no corporation or political entity has legal claim to nature.

We know there has been massive tree loss since some countries are making pledges to restore and conserve forested areas. Even by 1990, Serres (1995) says we were suffering harsh results and worse to come without careful thinking about future action. Though we experience life as individuals, we operate together with massive consequences. Compare the outcomes of human actions, environmentally degrading and polluting, with the ecological engineering of forest wildlife. Serres suggests as much by saying we have "invaded" the world's physical and ontological spheres (1995, p. 20). This is why we need above any contract based on political theories a contract with nature based in biology. However, each contract seems to depend on the other; we can't have just one. The problem is that judges are not in tune with the steady progression of scientific thought. Peter Burdon (2015) says the legal system must embrace eco-centric ethics and not see nature as an object of private property, a legal concept since Locke that has escalated environmental exploitation through hierarchy. Laws uphold the privileges of the powerful and wealthy to violate what environmental science declares we need for climate health. As Serres continues, politicians live for the short-term of election cycles, not for climate stability through long-term renewal and conservation (1995, p. 29).

People talk of private life exclusive of public law without considering how nature's wildlife owns and experiences a remote existence exclusive of public legislation. Serres (1995) might consider what's offered here as a battle between law (what could be) and bioecology (what is), a dialogue with no concrete resolution. There is "scientific reason" and yet "judicial error," he says (1995, p. 81). This human trial of tugging between extremes has little to do with the science of natural history but has much legal sway over dominion of forests. If most legislators were to become more environmentally and less economically inclined, perhaps the balance would tip in nature's favor. A problem centers on who has competency to speak on behalf of wildlife. In our world, authority (law) and not competency (science) rules. Serres goes on to suggest that whatever the outcome, a public trial can paradoxically offer science the "status of truth" (1995, p. 83). Yet, science and technology alone are not the solution without simultaneously boosting ecological ethics with cultural evolution.

Nevertheless, biological knowledge can help establish sovereignty for the forest state of Gaia. On the one hand, because of unfettered human activity and population explosion over the past few hundred years, there has been disordered global change. On the other hand, to correct the balance in justice to justify our past wrongs, we could use ordered global change that recognizes, respects, and is committed to forest sovereignty. Where law is judgment, science is reason; we need both working together, as Serres the philosopher declares (1995, p. 93). One way of achieving the goal of forest sovereignty and overcoming ignorance is by educating young people about the powers they hold in their abilities to act and elect responsibly, in line with how Serres claims that "intelligence adapts" (1995, p. 95). For Serres, there are relations with the world through a social and natural contract, a meeting of consequence. The tendency to pull together occurs in nature, from the structure of cells in organisms to forest ecosystems.

Ecological Ethics

We are destroying the environment fueled in part by warring ideologies of science and religion as well as separation of knowledge and action. Following Mary Midgley (2005), James Lovelock (2005) blames scientists for their reductionist methodologies that lean too heavily on reason and data. That appraisal might not be entirely true with biologists, according to Ernst Mayr (1997). As a superorganism, Gaia is more than theorems and equations centered only around humankind. With climate change out of control and planet earth in crisis, we've proved our incompetence. Midgley provides a socially relevant philosophy that does not stoop to materialism seeing earth as a machine to be engineered. She emphasizes free choice and personal responsibility in moral development, so individuals matter in her worldview. Biological science helps us understand how forests work and what makes them sick. Philosophy is just as important in helping us adjust ways of thinking to prevent further destruction and initiate forest preservation. A scientist might look at a map of the world and see all of the intricacies of ecosystems; a philosopher looking at the same map might see the character and values of societies in relation to those ecosystems. Both are structures of understanding what needs to be reconciled in order for a paradigm shift. We require an imaginative vision of nature to grasp earth's salvation. Science paints an accurate picture and makes predictions, but beyond the realm of science we have individual character and consciousness, group mentality, and societal values that could save or doom the planet.

Quantum physics tells us that on the smallest level molecules are in reference to each other, not separate. Some might see these relations as the mystery of life; but facts show that on every level of life there are living systems with population dynamics that sustain each other, even in a balance of creative destruction. The problem with modern humans is that we've tilted the scales of justice to force forms of life previously in creative relation with each other to compete destructively. As Midgley (2005, p. 19) might articulate, there's moral questioning of the individual relating to "origins in the natural world," whether a tree, beaver, or human. Rather than a vision of human life in the twenty-first century, people need to imagine what kind of natural world they want: burned and depleted or lush and fruited. According to Midgley (2005), recalling Hannah Arendt, the mechanistic view of "nature" as a separate repository of resources has led us down a path of unreflective annihilation. Instead, we must now see our world as Gaia, a revelation insisting on respect for nature. One reason we cannot see Gaia, suggests Midgley (2005, p. 356), is because our moral and political frames of mind have psychologically blocked us from understanding the fragile interconnectedness of nature. Rational and emotional understanding concerning the politics of nature come from a variety of philosophical and scientific sources, as I trust I've shown.

We might have finally broken the dualistic view of life (e.g., human/animal, society/nature) only because so many more people, cities, and nations are now subject to natural disasters, whether droughts, extreme heat, superstorms, or flooding. Nature, we now see, is not separated from us and is not static; we are from and live in the energies of nature. Our destruction of the planet cannot be reversed, and it might, at this point, not even be stalled. Even if all humans die, life will go on, and perhaps that is the ethical choice we have made. Enlarging and protecting forest areas with sovereignty and recognizing them as a transnational state of Gaia is a necessary yet complex first step in the right direction. Since animals are part of and indeed own a geography, there are ethical intonations involved. Human activity has serious consequences on their domains (Buller 2015). There can and should be property ownership of place for organisms large and small who have a "vital interest" in their goods of nature (Hadley 2005). W. D. Ross (1998) talks about the rightness of acts and the goodness of motivations. While it's difficult to assign right and good to forest creatures, elements of our own morality evolved in the forest (Tague 2016, ch. 2). Therefore, from an anthropomorphic perspective we could say what's right and good are intuitively apprehended by forest organisms in their duties as ecological engineers. Even if for their personal benefit, the combined results of organisms are mutually advantageous in the forest garden.

Paleoanthropologists teach that we apes evolved along with and simultaneously to many other forest dwelling primates. Separation from and destruction of our kin forests seems unethical. From a practical standpoint, we still need forests for life since they constitute part of a family tree from which we are closely allied. There are moral questions in our flattening of forests that hold our roots and future livelihood. Forest Gaia is at once independent of us and yet also in our evolutionary history just as the air we breathe. As Midgley (2005, p. 368) would say, forest ecology is more important than human economics. The moral dilemma here is that most people, and specifically world leaders and politicians, place inward-facing economics over outward-looking ecology, thus explaining why over the past 50 years or so the climate crisis has intensified.

Ecological ethics, to use Patrick Curry's (2011) phrase, homes in on eco-centric principles related to how one acts in and treats the natural world. Our ethos and behavior are germane to saving forests. This means that human values are essentially derived from and should become embedded in nature, not in corporations, political institutions, or technology. The problem, as Curry seems to suggest, is that anthropocentric ethics has totally replaced and nearly eliminated eco-centric values. Human-only morality has served its stilted purpose well for those who have profited handsomely in using nature to make and sell disposable items, but a shift in ethics would turn to valuing forests, a rational priority to support all life forms. Recalling Aldo Leopold and E. O. Wilson and following Hannah Arendt and Mary Midgley, Curry's ecological ethics is not wholly geared toward scientific reasoning's truisms but depends on emotional connections to earth and standards of fairness and justice (2011, p. 26). Some see salvation in religion, others in science, but neither on its own will save earth. Thomas Berry (2014) has made an effort to blend the two in what he calls the Ecozoic Era, where we mutually cooperate with nature in an earth community. Attitude changes via cultural evolution are necessary as we look to green our values. Using social media as a survey base, researchers found that mostly in developed countries there's increasing awareness of and concern about climate change (Leiserowitz, et al. 2023). Anthropocentrically, the report makes no mention of wildlife or forests. So, there's room for moral change on several fronts.

Deontology or rights derive from Kant's categorical imperative, which states that one must rationally act as if following a universal maxim treating people as ends and not as means, as subjects and not as objects (Kuehn 2001, p. 286). Deontological Kantian ethics is a rational moral imperative not tied to consequences or other contingencies. Autonomous self-governance of the willful and reasonable individual is paramount in Kantian ethics so that no policy should exploit or sacrifice the worth of an individual for some goal outside the scope of the individual's way of life (Christman 2002). Rawls embraces this thinking.

Plants and animals have been excluded from this philosophy, though Kant realized that animal cruelty reveals character corruption and fosters harms to people. Implied is the question of how one expects others to treat her. Shift your thinking and ask how the forest state of Gaia has treated us. Kant, unlike Hume, did not dwell on sympathy, yet values about nature are not merely rational but emotional.

Jeff Sebo (2023) says we can have "fundamental values" that change in certain contexts. There are, he believes, different Kantian duties from humans over mice over ants (his example). Imagining, but not fully condoning Kantianism for humans and utilitarianism for nonhumans, Sebo claims duty depends on individual characteristics and relationships, "choice situations." While there's merit in grappling with these definitions and applications, the result is confusing. As another path, virtue ethics depends on one's ethos and motives, a social character of temperance and self-control, honesty and integrity, fairness and intelligence. The aim of virtue ethics is to live well enough so as to thrive and do good for others. This way of life often happens not by reason or calculation alone but is accompanied by emotional understanding and sympathy, kindred feeling. Individual flourishing is the moral good and could be measured, considering social contexts, by how any state helps the individual achieve the ideal life, a view that can be seen as communitarian and anti-liberal (Christman 2002). Much of my ecologically ethical argument is based on how as industrialized people most of us have lost any fellowship with the natural world.

The intrinsic and instrumental values of a forest are intertwined since they exist for the benefit of all life forms. There has been much talk in the past decade or so about sustainability, but that stance, like environmental impact statements, seems more for the advantage of humans than for preserving isolated regions of forests, wetlands, or grasslands. Evidenced in how many business schools and management programs now fly under the banner of sustainability, the movement is a human enterprise driven by the utilitarian rule of political-corporate forces. Curry (2011, p. 67) says as much and cites sources to verify how corporations speak of sustainability to promote business as usual. The intrinsic value of forest flora and fauna are subjugated to the instrumentality of "sustainability." One example Curry uses (2011, p. 211) are wind farms, which he labels as "factories." These ventures are funded by large corporations and investment banks for financial gain. Their rationale is not to lessen energy usage but to pump it up for greater profit. Returning to ideas outlined in the beginning, we have yet to establish in line with Rockström and Klum (2018) proper ecological boundaries.

Ecological Thinking

Some might accuse me of raising an anthropocentric ethics by claiming we need forests for human life. This is untrue since forest health is necessary for the survival of all life forms. A report from the IPCC (Lee 2023) highlights necessary land use changes regarding reforestation and restorative management of wetlands and grasslands. Generally, my thinking holds an underlying assumption of intrinsic value for every living organism. Typically, humans compete unethically against nature. What is needed and called for (Curry 2011, pp. 110–11) is ecological thinking and activism that fosters conservation and preservation, social, political, and institutional change to control population growth and limit mass consumption and waste. These positive movements can occur by honestly educating the young with politicians who are committed to a cultural evolution that will prevent the living environment from perishing. Much ideological broadcasting, on the other hand, is used to attack vulnerable people, the environment, and animals.

As Curry (2011, p. 112) might say, we've reached a point where one must give legitimate reasons for wholesale deforestation. One can't say jobs are saved by tearing down a mountain for coal. Rather, we need to situate human issues in the larger ecosphere. Though one might assert the obvious need for an eco-centric mindset assuming the universal truth of Gaia, because of philosophical pluralism others will disagree and posit many alternative views. The various political philosophers outlined in this book prove the power of multiple truths. However, anthropocentric ideology, which many philosophies embrace, has failed the natural world to which we belong. Without clean air or water, no one survives; but we can imagine how with the demise of human life other creatures will persist and eventually flourish. Curry (2011, p. 161) talks about the need for a green citizenship where one is much less a consumer and more a leader of ecological ethics. Citizens can speak out, raise political issues, and vote. One activist who comes to mind is Jane Goodall. There's also David Attenborough, cited earlier. Another prominent voice belongs to Greta Thunberg. These figures wield tremendous cultural influence, though some on the politically conservative spectrum have unethically demonized Thunberg and environmentalists like her.

If one can see the intrinsic value in nature around her home, maybe she can understand why eco-centric values are important globally. Over many millennia, species have succumbed to Darwinian change, but forest Gaia has persisted efficiently. Observation reveals that the main threat to wild flora, fauna, and fungi is human curtailment of feeding areas and pressure from human-induced climate change. Forest organisms emit sustainable waste that restores the biotic community. Humans, rather, create toxic chemical wastes that infect wildlife. The transnational state of Gaia is more ecologically ethical than most industrial societies, though pressure from conservationists on consumerist countries is having a discernable public relations effect, albeit minor. Increased populations will only create among industrial nations more consumerism and dirty waste with the need for fossil fuel energy. We are literally overpopulating and outpacing land use, a resounding death knell.

From evidence he acquired living and researching in Ecuador's Amazon region, Eduardo Kohn (2013) says people should develop ways of living with other creatures in nature. There are possibilities of new relationships and understanding. We need to rethink, he suggests, how we represent nature's wildlife and forests in our consciousness. We are not exceptionally raised above and separate from the natural world, yet we need to acknowledge the "important distinctions" (Kohn 2013, p. 7) between humans and animals. In any nonhuman anthropology, the world is not represented by human language. As far as we know, for wildlife the forest is embodied in and represented by visual images and colors, light and darkness, wet and dry, hard and soft, sounds and vibrations (even silence), aromas, scents, and odors, textures smooth and coarse, and a multitude of tastes, a gestalt of simultaneity in any creature's Umwelt.

I should be careful since this description points to a dualistic interpretation. Kohn (2013, p. 7) would remind us that while we don't want to "flatten important distinctions," there's some Darwinian continuity across species' inner lives. Comprehending and accepting these physical and emotional connections will enable us to live with wildlife and not destroy it. To emphasize Kohn's point (2013, p. 16), ontological reality is not only human but an "ecology of selves." In short, Kohn wants to decolonize anthropocentric thinking about forest flora, fauna, and fungi. That initiative is an ethical move. As he asserts, forests think on their own. Perhaps overly stretching Kohn's meaning, modern humans establish forest presence by creating its absence and the elimination of signs denoting the wild. For industrialized nations employing corporate powers this erasure is not Darwinian evolution, not natural to an ecosystem or ecological niche construction. There is, rather, separation without an emergent human-forest ethos.

How we see the natural world from a wildlife perspective should matter. Ecosystems are collections of organisms and soils responding to energies in and forces against existence in complex and mutual relationships for stability in a bio-network of common good. As a person of his time Kant viewed animals in service to humans, but he was quite clear on the duty of care we have toward nonhuman life forms. In a close paraphrase from his *Metaphysics of Morals* (1996, pp. 192–93), Kant says that the destruction of what is beautiful in nature weakens the human disposition that promotes morality. Moreover, cruelty or violence toward animals erodes one's sense of morality and sensitivity to another's pain. In a similar vein, drawing from Hinduism, Daoism, and the Confucian traditions, empathic positions see the suffering of all in one or the forest in each tree. The human world relies less on physical sensations and more on abstract thinking. While Neil Evernden (1993) suggests we have mostly alienated ourselves from the intricacies of the natural world, let's go further and say how this separation has resulted in human minds and bodies acting as a virus to nature. Humans, unlike most organisms, have no definable natural niche but create a modern social world that has exploited nature. The key question is how we will shift away from consumerism that damages forests so that we preserve the vital life-enhancing powers of wildlife and plants.

Our fate need not be one of doom and pessimism but a cultural turn to perceiving nature as our *alma mater*, not a backdrop to our self-centered consciousness, needs, and desires. Formulated by a range of scientists and philosophers, The New York Declaration on Animal Consciousness (NYU 2024) shows that science is producing evidence demanding the attention of politicians and policy makers. Darwin (2003, ch. 14) submits that an organism is an evolving form of continuous change and adaptation to circumstances through modified descent with mutual relations among other organisms. We can see Gaia similarly where the biological fate of one species is entangled with many others. Taking into account how flora, fauna, and fungi constitute a single superorganism as per James Lovelock and E. O. Wilson, any probability of the consciousness of Gaia should demand its protection.

Final Word: Indigenous Voices

Indigenous people don't separate human/nature, and for them the Western word "nature" carries meanings and implications far different than what we've seen in the political philosophy of Chapter 2. Contrary to Rousseau, Indigenous people are not magically connected to nature; their reciprocal relationship with nature is practical and spiritual. Close to how Lovelock and Margulis present Gaia, Indigenous people see the natural world "holistically as a living system" reflected in their practices and ceremonies (Andersson, et al. 2021, p. 3). Indigenous people don't marginalize nature as do their colonizers. To this day, many national parks were once prime lands for Indigenous people who, from Canada and Finland to the Amazon and Pacific Rim, are good managers of nature because they are aware of their role in Gaia's systems. Joshua Reid (2021) says that post-colonial attitudes create policies of national parks meant to exclude Indigenous people, obviously what's not recommended in this book. National parks are protected spaces created by Western political philosophy in "green colonialism" (Klein 2016) that unethically separates Indigenous people from their homelands while constructing a capitalistic semantics of nature, wilderness, and preservation,

Reid suggests. Readers should see by now that I do not conflate the colonist, settler mentality of parklands with a forest state of Gaia. By seizing native lands and ignoring rights, contemporary states still colonize, and often both secular and religious moralizing undermine the legitimacy of Indigenous views, laws, and land management.

There are legitimate legal disputes pursuant to prior consultation with Indigenous people regarding natural resource rights in, for instance, Colombia. Economic development like dam building in corrupt countries, says Cesar Rodriguez Garavito (2011), with paramilitary governments and armed guerillas in native lands immorally takes precedent over Indigenous rights to hunt and fish sustainably. Fractious uprisings show the political and social neglect of Indigenous people, like the Peruvian Andeans and Amazonians. Where lands should rightfully be decreed to natives, politicians are permitting capitalist predators with financial interests to privatize and buy Indigenous lands that sustain the resources of Gaia. Actions like these, whether from democratic or socialist countries, erase in one stroke Indigenous rights and their autonomy in Gaia, says Gerardo Rénique (2009). Legal recourse, as outlined in this book whether in the form of a revised reading of political philosophy or community actions, is necessary to preserve the self-determining aspects of Gaia represented in flora, fauna, and Indigenous people. In this instance and without any concessions, Rénique continues (2009, p. 123), the leaders of Peru, namely President Alan García, publicly said that the Indigenous people had no rights to land and should be replaced by the bourgeoise and their GDP-generating technology. (In 2022 there was a proposal in Peru to strip Indigenous peoples of their protections.) This capitalistic mentality, says Rénique, creates oppositions of plunder and resistance, not a way to preserve Gaia or move forward in ethical politics, whether from Rawls, Nozick, or Walzer.

As Rani-Henrik Andersson et al. point out (2021, p. 7), co-management of ancient forests by Indigenous and modern people has not benefited the true inheritors of the land since policy makers and environmentalists often restrict traditional hunting and fishing. Andersson et al. highlight successful cases of management, so there is hope for industrial humans in Gaia with the assistance of native people's wisdom. In the early seventeenth century, Native Americans helped immigrant Pilgrims; as years passed, this act of generosity was subsequently repaid with genocide. Western political philosophy epitomized in Locke has not been a boon to Gaia, but turning to Indigenous perceptions and practices is a way forward with protectionist policies flowing from a new reading of philosophers as presented here. Going well beyond the Lockean notion of one possessing land he has confiscated for cultivation or development, Laura Pulido (2017) says such appropriation is an act of environmental and social capitalism with violence sanctioned by a nation-state. Indigenous people don't romanticize nature as "out there" but build sacred and practical relationships with biotic and abiotic aspects of nature, worlds both visible and invisible, whether a fallen tree in a coppice or a beaver dam in a river (Andersson, et al. 2021, p.8). Unfortunately, many Native Americans and similar people worldwide have been dispossessed of their lands and thus culturally and economically deprived. That's the tragic human loss, and we see deprivation parallels in how industrialized people using political philosophy as a crutch have deprived Gaia of its vitality, too. As mentioned previously but worth emphasizing, any policy decision about the preservation of the forest state of Gaia must include Indigenous voices. Not to generalize, with Indigenous people (whether in North or South America, Africa, Indonesia, Russia, Finland, or Sweden, etc.), a culture of forest care and respect for plant and animal life could be critical ingredients to protecting the forests of Gaia since the health of the local is intimately tied to global well-being.

According to Bruce Clarke (2020, p. 60), rather than participating in nature as do Indigenous people, industrialized nations have habitually alienated themselves from Gaia. Worse, modern, industrial attitudes reflecting capitalistic gains have dehumanized Indigenous people as "others," and some corporations or nation-states are cynically acquiring lands as carbon offsets (Klein 2016). If eradicating the customs of Indigenous people is cultural genocide, consider as well the fading storehouse of environmental behaviors of all species in Gaia. The issue is that in the viral war against Gaia for the past few hundred years, all economic rewards and privileges flow to rich nations while the weight of toxic risks and sacrifices have been pushed onto Indigenous people and forest wildlife. Though idealistic, in order to generate a healthy planet marginalized people reinforced by global youth could demand a cultural revolution to eliminate the inequalities against other people and the forest state of Gaia. As one of many Indigenous representations of a Gaian worldview, let's turn to the Yanomami of the Amazon rainforest in northern Brazil and southern Venezuela. In a time of climate crisis and biodiversity loss, to whose voices should reasonable people defer? To corporate executives in gleaming towers? Or to Indigenous people who see their communal relationship with plants and animals as interactive in "multispecies societies" (Anderson 2024, p. 9). Indigenous voices represent nonhuman species often forgotten or viewed as commodities, disposable objects obstructing the path to commercial production.

There are rights of nature, and examples can be found as follows (and easily searchable online): Personhood for Lake Erie, U.S.; Personhood for the Klamath River by the Yurok Tribe, California; The Whanganui River, New Zealand; the Atrato River, Colombia; the Yamuna and Ganges Rivers, India; the White Oak

Tree that owns itself, Georgia, U.S.; Taranani Maunga Mountain, New Zealand; Te Urewera lakes and forest areas, New Zealand. Indigenous and native people and their representatives are responsible for conservation that recognizes the intrinsic worth of nature. Bradshaw (2020) notes how Indigenous people are forerunners in the area of granting rights to nature. Industrial communities both urban and rural across the globe could learn from this traditional respect for the natural world.

Indigenous people like the Yanomami don't see separate species but traces in everyone, a "multispecies cosmopolitics" in contrast to how most nation-states are predicated on some of the political theory covered in Chapters 2 and 3 (Anderson 2024, p. 22). Providing the groundwork to much Western philosophy, Hobbes is reductive; relationships are between a victim and victor following what he calls the laws of nature. While not as pessimistic as Hobbes, Locke favors human reason and authority over the rawness of natural equality. Building from Hobbes and Locke, Rousseau does not see anarchy in nature but the basis of a social contract. For Hobbes there's law by imposition and constraints; for Rousseau, who draws from Montesquieu, there's "popular sovereignty" enabled by natural interrelations and associations (Anderson 2024, p.169). Anglo-European thought tends toward the subjective individual influenced by patriarchy in nation-state lands, unlike Indigenous views of a collective with holism and forest state ecosystems. For instance, as Mark Anderson (2024, p. 172) points out, with the Yanomami there are no words for land or forest, and a place without trees is considered dead. The Amazonian Indigenous people don't settle but move to different locations periodically so that plants and animals can recover. There's little to no acquisitive accumulation (see Karl Marx) contrary to Western financial practices (see Adam Smith). The Yanomami have no sovereign but rather a decentralized, communal panel that works toward consensus, not dicta (see the proposal in the Preface here).

Drawing from Kohn (2013), Anderson (2024, p. 21) explores Yanomami biosemantic interactions among rainforest species in contrast to the political theory of Hobbes and others who elevate the human person. There's a multispecies government of plants and animals who relate in an idiom we cannot grasp. Self-awareness is not linguistic. Their politics is one of diplomacy, not supremacy. In tandem, for the Yanomami, community communication can be affective song and not just discrete words, a politics that embraces and does not exclude the multispecies sounds of the natural world. With Indigenous thinking, plants and animals are fundamental to, and open a way for, all ethical and any political relationships. In fact, says Anderson (2024, p. 21), the Yanomami see politics as the "state of plants" where all relations are open and oriented to flora, a collaboration established by differences. The ecosystem is not property that belongs to someone or a group but to all inhabitants in shared responsibility. From most modern human perspectives, plants are seen as passive, but this is not so for the Yanomami. An ethics of plants is a nexus to form multispecies distributions and relationships (Anderson 2024, p. 189). Affirming life, plants avail themselves as nourishment for others; the total being of plants can be traced in other elements of life and even in rocks, stones, rivers, the sky, etc. Plants give without seeking exchange, as represented in the Yanomami and distinct from the philosophy we've seen in Chapters 2 and 3 (Anderson, p. 197).

Anderson (2024) notes how *The Falling Sky* by shaman Davi Kopenawa highlights the dishonesty of political discourse in contrast to the truth of species vernacular. For Kopenawa and the Indigenous people of Yanomami, politics is not transactional negotiation but hospitality, a diplomacy of song among and between humans and nonhumans promoting friendly relations. For the Yanomami, and likely other Indigenous people, mirrored in the work of philosopher Michael Marder, there's an ethical interface as one confronts the plant world, the other that one can or not ethically accommodate (Anderson 2024, pp. 189–90). Many Westerners find this natural interaction difficult to understand or accept. As anthropologist Bruce Albert (Kopenawa & Albert 2013) says in *The Falling Sky*, after a long fight and personal loss, at the Earth Summit at Rio de Janeiro in 1992, Davi Kopenawa obtained from Brazil legal recognition of Terra Indígena Yanomami, a large tropical area. The fight was over land taken and destroyed by road construction, cattle farming, and gold mining. Kopenawa continues to challenge the "arrogant deafness" of white people leading a fight against the "People of Merchandise" decimating the forest home of his people (Kopenawa & Albert 2013, pp. 5, 8). The book is a testament to aboriginal voices, so let's report some of what shaman Kopenawa shares in a very lyrical manner.

Kopenawa laments the limitations of language in the defense of forest land as opposed to the shaman's words that resonate inside of people. Affirming some themes of this book, he says that the minds of great white men, on the other hand, are "full of oblivion" as they spread literal and figurative epidemics against people and the fertility of trees while speaking mendaciously about politics (Kopenawa & Albert 2013, pp. 307, 312). Kopenawa traveled to London, Paris, and New York. At that time, he was dismayed by the sheer amount of noise, concrete, and material goods of people who pretended to love nature. He's aware that these modern Western countries destroyed their own forests for the love of merchandise, replacing woodlands with buildings and factories to manufacture yet more goods. The Westerners, he sees, are enchanted by machines and material possessions with "limitless desire" (p. 327). The Yanomami, on the other hand, generously barter goods without waste or greed. For him and the Yanomami, wants are simple and always rewarded: a silent forest and a clear sky. The Western city dwellers, he suggests, in their epidemic desire for material goods have collectively lost their memory of and friendship for the forest and seize biodiverse land not for beauty but to clear trees for livestock, cattle feed, and mines (p. 358).

In a key chapter entitled "The Spirit of the Forest," Kopenawa almost anticipates the climate crisis in terminology Lovelock and Margulis might advocate by saying that mechanistic forest destruction will cause the sky to grow dark, fall, and crash to earth with baneful effects (Kopenawa & Albert 2013, p. 408). Westerners destroy what they don't care to know, which they think is dead already, clearing and killing land "without measure" (p. 382). His people, rather, have knowledge of forests and plants, and in spite of subsistence hunting they designate animals as "fellow creatures" (p. 387) comprehending that all living beings come from the same spiritual forest root. Echoing motifs of this book regarding Gaia, Kopenawa says that those nonhumans who live in the forest are masters of time, owners who affect air and winds, rainfall and rivers, and seasons in constant renewal (p. 389). What Westerners singularly call nature, and by using fancy words like environment or ecology, only translates to destruction for timber, cattle, or gold. Central to the claim of this book, the Yanomami understand that earth "cannot be split apart" as if the forest is an afterthought when it's truly part of a nourishing system (p. 397).

Let's conclude with some words from Emanuele Coccia, who comments on the work of Kopenawa and Albert. Again, resonating ideas from this book, Coccia (2022) says the collaboration of Albert and Kopenawa has initiated a post-anthropological ethnography that no longer distinguishes human from nonhuman ecology (Coccia 2022, p. 17). Coccia points to how some very different cultural voices try to find unison. Is that possible in today's political climate of division? Ecology, he says, because of *The Falling Sky* is not an abstraction to be accomplished somewhere over there but is a present claim on living beings in how people and animals are not so distinct. This idea would not surprise the Yanomami or other Indigenous people but contradicts Judeo-Christian beliefs. Drawing from anthropologist Claude Lévi-Strauss, Coccia bemoans how the Western tradition has lost the ability to commune and communicate with the plant and animal world, thus losing its own identity. That's true as I've outlined, but there are hopeful models, like Saint Francis of Assisi. One can pray that *Forest Sovereignty*, along with the many works it draws from, will point us to resurrecting our symmetry with wild plants and animals.

Bibliography

Alexander, Richard D. 1987. *The Biology of Moral Systems*. NY: Aldine de Gruyter.

Allen, Colin. 2001. "Cognitive Relatives and Moral Relations." *Great Apes and Humans: The Ethics of Coexistence*. Benjamin B. Beck, et al., eds. Washington, DC: Smithsonian Institution Press. 261–273.

Almond, R. E. A., et al. 2022. *Living Planet Report 2022*. WWF, Gland, Switzerland.

Aminetzah, Daniel, et al. 2022. *Nature in the Balance: What Companies Can Do to Restore Natural Capital*. NY: McKinsey & Co.

Ammann, Karl. 2001. "Bushmeat Hunting and the Great Apes." *Great Apes and Humans: The Ethics of Coexistence*. Benjamin B. Beck, et al., eds. Washington, DC: Smithsonian Institution Press. 71–85.

Anderson, Mark. 2024. *The Rights of Nature and the Testimony of Things*. Nashville: Vanderbilt UP.

Anderson, Mark G. and Charles E. Ferree. 2010. "Conserving the Stage: Climate Change and the Geophysical Underpinning of Species Diversity." *Plos One* 5 (7): e11554. <https://doi.org/10.1371/journal.pone.0011554>

Andersson, Rani-Henrik, et al. 2021. "Traditional Indigenous Knowledge and Nature Protection." *Bridging Cultural Concepts of Nature: Indigenous People and Protected Spaces of Nature*. Boyd Cothran and Saara Kekki, eds. Helsinki: Helsinki UP. 1–25.

Andreoni, Manuela. 2024. "Can Forests Be More Profitable Than Beef?" *The New York Times*, 2 May. <https://www.nytimes.com/2024/05/02/climate/amazon-reforestation.html>

Andrews, Kristin and S. Monsó. 2021. "Animal Cognition." *Stanford Encyclopedia of Philosophy*. <https://plato.stanford.edu/entries/cognition-animal/>

Andrews, Kristin, et al. 2024. "Human and Nonhuman Norms: A Dimensional Framework." *Philosophical Transactions of the Royal Society B* 379: 20230026. <https://doi.org/10.1098/rstb.2023.0026>

Arendt, Hannah. 1998. The Human Condition. Second edition. Chicago: U Chicago P. (Original work published 1958)

Aristotle. 1952. "Nicomachean Ethics." *The Works of Aristotle*, volume 2. Robert Hutchins, ed. Chicago: Britannica.

Attenborough, David. 2020. *A Life on Our Planet*. NY: Grand Central.

Augustine, Aby, et al. 2022. "A Study on the Greenwashing Strategies in BASF's 'We Love EVS' Campaign." *Kristu Jayanti Journal of Humanities and Social Sciences* 2: 56–64. <https://doi.org/10.59176/kjhss.v2i0.2224>

Bakker, Peter, et al. 2021. *Vision 2050: Time to Transform*. World Business Council for Sustainable Development.

Barry, Joshua M., et al. 2018. "Pumas as Ecosystem Engineers: Ungulate Carcasses Support Beetle Assemblages in the Greater Yellowstone Ecosystem." *Oecologia* 188. <https://doi.org/10.1007/s00442-018-4315-z>

BASF. 2023. Examples of Biodiversity Products. <https://www.basf.com/us/en/who-we-are/sustainability/we-produce-safely-and-efficiently/resources-and-ecosystems/biodiversity-and-ecosystems/examples-of-specific-projects.html>

Bastin, Jean-Francois, et al. 2019. "The Global Tree Restoration Potential." *Science* 365 (6448): 76–78.

Baxter, Brian. 2005. *A Theory of Ecological Justice*. London: Routledge.

Benyus, Janine, et al. 2022. "Ecological Performance Standards for Regenerative Urban Design." *Sustainability Science*. <https://doi.org/10.1007/s11625-022-01180-5>

Bekoff, Marc and Jessica Pierce. 2009. *Wild Justice: The Moral Lives of Animals*. Chicago: U Chicago P.

Berg, Bård A. 1996. "Government Intervention Into Sámi Reindeer-management in Norway: Has It Prevented or Provoked 'Tragedies of the Commons'?" *Acta Borealia* 13 (2): 69–89. <https://doi.org/10.1080/08003839608580454>

Bergson, Henri. 1911. *Creative Evolution*. Arthur Mitchell, trans. Lanham, MD: University Press of America.

Berry, Thomas. 2014. *Selected Writings on the Earth Community*. Mary Evelyn Tucker and John Grim, eds. NY: Orbis Books.

Bianchi, Carlos A. and Susan M. Haig. 2013. "Deforestation Trends of Tropical Dry Forests in Central Brazil." *Biotropica* 45 (3): 395–400. <https://doi.org/10.1111.btp.12010>

Blackstone, William. 1979. *Commentaries on the Laws of England*. Chicago: U Chicago P.

Blattner, Charlotte, et al. 2020. "Animal Agency in Community: A Political Multispecies Ethnography of VINE Sanctuary." *Politics and Animals* 6: 1–22.

Boch, Steffen, et al. 2013. "High Plant Species Richness Indicates Management-related Disturbances Rather Than the Conservation Status of Forests." *Basic and Applied Ecology* 14: 496–505. <https://doi.org/10.1016/j.baae.2013.06.001>

Bochaton, Corentin, et al. 2021. "Large-scale Reptile Extinctions Following European Colonization of the Guadeloupe Islands." *Science Advances* 7:eabg2111. <https://doi.org/10.1126/sciadv.abg2111>

Booth, Annie L. 2003. "We Are the Land: Native American Views of Nature." *Nature Across Cultures*. Helaine Selin, ed. Kluwer: Dordrecht. 329–349.

Boucher, Doug, et al. 2012. "Grade A Choice? Solutions for Deforestation-free Meat." *Union of Concerned Scientists*. <www.ucsusa.org/deforestationfree>

Bourne, Edward Gaylord. 1901. *The Federalist*. Washington: M. Walter Dunne.

Bradshaw, Karen. 2020. *Wildlife as Property Owners: A New Conception of Animal Rights*. Chicago: U Chicago P.

Brakes, Philippa, et al. 2021. "A Deepening Understanding of Animal Culture Suggests Lessons for Conservation." *Proceedings of the Royal Society B* 288: 20202718. <https://doi.org/10.1098/rspb.2020.2718>

Brehm, Allison M., et al. 2019. "Land-use Change and the Ecological Consequences of Personality in Small Mammals." *Ecology Letters* 22 (9): 1387–1395. <https://doi.org/10.1111/ele.13324>

Brewbaker, J. L. and W. G. Sun. 1999. "Trees and Heterosis." *Genetics and Exploitation of Heterosis in Crops*. James G. Coors and Shivaji Pandey, eds. Madison, WI: American Society of Agronomy. 463–477.

Brienen, R. J. W., et al. 2018. "Long-term Decline of the Amazon Carbon Sink." *Nature* 519 (7543): 344–348. <https://doi.org/10.1038/nature14283>

Browne, Janet. 1995. *Charles Darwin: Voyaging*. Princeton: PUP.

Buckley, Cara. 2023. "The Giving Forest." *The New York Times*, 22 April.< https://www.nytimes.com/2023/04/22/climate/menominee-forest-sustainable-earth-day.html>

Buller, Henry J. 2015. "Animal Geographies III: Ethics." *Progress in Human Geography*. University of Exeter, England. <http://hdl.handle.net/1087/17084>

Bump, Joseph, et al. 2022. "Predator Personalities Alter Ecosystems." *Frontiers in Ecology and the Environment* 275–276. <https://doi.org/10.1002/fee.2512>

Burdon, Peter D. 2015. *Earth Jurisprudence: Private Property and The Environment*. Abingdon, Oxon: Routledge.

Burkhardt, Helmut and Richard Westra. 2008. Introduction to Part IV. *Reconciling Human Existence with Ecological Integrity*. Laura Westra, et al., eds. London: Earthscan. 215–217.

Butynski, Thomas M. 2001. "Africa's Great Apes." *Great Apes and Humans: The Ethics of Coexistence*. Benjamin B. Beck, et al., eds. Washington, DC: Smithsonian Institution Press. 3–56.

Calvo, Paco with Natalie Lawrence. 2022. *Planta Sapiens: The New Science of Plant Intelligence*. NY: W. W. Norton.

Camia, A., et al. 2021. *The Use of Woody Biomass for Energy Purposes in the EU*. EUR 30548, EN, Luxembourg. <https://doi.org/10.2760/831621>

Carson, Rachel. 2002. *Silent Spring*. Boston: Mariner Books. (Original work published 1962)

Cartwright, Paulyn. 2014. "Origins and Early Evolution of Animals." *The Princeton Guide to Evolution*. Jonathan B. Losos, ed. in chief. Princeton: PUP. 159–166.

Center for Sustainable Systems, University of Michigan. 2021. "Carbon Footprint Factsheet." Publication number CSS09–05.

Christman, John. 2002. *Social and Political Philosophy*. London: Routledge.

Clapp, James Gordon. 1967. "John Locke." *Encyclopedia of Philosophy*. Paul Edwards, ed. NY: Macmillan. Volume 4: 487–503.

Clark, Michael A., et al. 2020. "Global Food System Emissions Could Preclude Achieving the 1.50 and 20 C Climate Change Targets." *Science* 370 (6715): 705–708. <https://doi.org/10.1126/science.aba7357>

Clarke, Bruce. 2020. *Gaian Systems: Lynn Margulis, Neocybernetics, and the End of the Anthropocene*. Minneapolis, MN: U Minnesota P.

Clement, Grace. 2003. "The Ethic of Care and the Problem of Wild Animals." *Between the Species* 13 (3). <https://doi.org/10.15368/bts.2003v13r3.2>

Cleveland, Amanda C. 2014. *Symbiosis Between Biophilic Design and Restorative Healing Environments*. Florida State University, MFA thesis.

Coccia, Emanuele. 2019. *The Life of Plants: A Metaphysics of Mixture*. Dylan J. Montanari, trans. Cambridge, UK: Polity Press. (Original work published 2017)

Coccia, Emanuele. 2022. Préface. *Yanomami, L'Esprit De La Forêt*. Bruce Albert and Davi Kopenawa. Paris: Foundation Cartier/Actes Sud. 15–22.

Cochrane, Alasdair, et al. 2016. "Animal Ethics and the Political." *Critical Review of International Social and Political Philosophy* 21 (2): 261–277. <https://doi.org/10.1080/13698230.2016.1194583>

Collini, Stefan. 1989. Introduction. *John Stuart Mill: On Liberty and Other Writings*. Cambridge: CUP. vii–xvii.

Cooley, Mackenzie. 2022. *The Perfection of Nature: Animals, Breeding, and Race in the Renaissance*. U Chicago P.

Cornford, Francis MacDonald. 1975. Introduction. *The Republic of Plato*. London: Oxford UP. (Original work published 1941)

Costa, James T. 2017. *Darwin's Backyard*. NY: W. W. Norton.

Costa, James T. and Bobbi Angell. 2023. *Darwin and the Art of Botany: Observations on the Curious World of Plants*. Portland: Timber Press.

Costanza, Robert, et al. 1997. *An Introduction to Ecological Economics*. St. Lucie Press and ISEE. E-book.

Coulter, Kendra. 2023. *Defending Animals: Finding Hope on the Front Lines of Animal Protection*. Cambridge, MA: MIT Press.

Crabbé, P. J. and J. P. Manno. 2008. "Ecological Integrity as an Emergent Global Public Good." *Reconciling Human Existence with Ecological Integrity*. Laura Westra, et al., eds. London: Earthscan. 73–84.

Crane, Peter R. and Andrew B. Leslie. 2014. "Major Events in the Evolution of Land Plants." *The Princeton Guide to Evolution*. Jonathan B. Losos, ed. in chief. Princeton: PUP. 143–151.

Crary, Alice and Lori Gruen. 2022. *Animal Crisis: A New Critical Theory*. Cambridge, UK: Polity Press.

Crowther, T. W., et al. 2015. "Mapping Tree Density at a Global Scale." *Nature* 525: 201–205. <https://doi.org/10.1038/nature/4967>

Cudworth, Erika. 2025. "Resisting Zoopolis: Bordering Species Relations as a Response to COVID-19." *Human-Animal Relationships in Times of Pandemic and Climate Crisis*. Josephine Browne and Zoei Sutton, eds. Abingdon, Oxon: Routledge. 133–150.

Curry, Patrick. 2011. *Ecological Ethics: An Introduction*. Second ed. Malden, MA: Polity Press.

Darimont, Chris T., et al. 2023. "Humanity's Diverse Predatory Niche and Its Ecological Consequences." *Communications Biology* 6: 609. <https://doi.org/10.1038/s42003-023-04940-w>

Darwin, Charles. 1989. *The Formation of Vegetable Mould Through the Action of Worms With Observations on Their Habits*. NY: NYU P. (Original work published 1881)

Darwin, Charles. 2003. *On the Origin of Species*. First edition. Joseph Carroll, ed. Canada: Broadview. (Original work published 1859)

Davenport, Coral. 2023. "U.S. to Fund a $1.2 Billion Effort to Vacuum Greenhouse Gases From the Sky." *The New York Times*, 11 August. <https://www.nytimes.com/2023/08/11/climate/carbon-dioxide-direct-capture.html?>

de Beer, Gavin. 1965. *Charles Darwin: A Scientific Biography*. NY: Anchor Books.

Del Complex. 2023. "BlueSea Frontier Compute Cluster: The Future of AI Innovation." <https://www.delcomplex.com/blue-sea-frontier>

de Waal, Frans. 1982. *Chimpanzee Politics: Power and Sex Among Apes*. NY: Harper & Row.

Ding, Jingyi and David J. Eldridge. 2024. "Trees Support Functional Soils in a Dryland Agricultural Area." *Geography and Sustainability* 5: 588–596. <https://doi.org/10.1016/j.geosus.2024.08.001>

Di Sacco, Alice, et al. 2020. "Ten Golden Rules for Reforestation to Optimize Carbon Sequestration, Biodiversity Recovery and Livelihood Benefits." *Global Change Biology* 27: 1328–1348. <https://doi.org/10.1111/gcb.15498>

Doherty, Thomas J. and Susan Clayton. 2011. "The Psychological Impacts of Global Climate Change." *American Psychologist* 66 (4): 265–276. <https://doi.org/10.1037/a0023141>

Donaldson, Sue and Will Kymlicka. 2011. *Zoopolis: A Political Theory of Animal Rights*. Oxford: OUP.

Donaldson, Sue and Will Kymlicka. 2013. "A Defense of Animal Citizens and Sovereigns." *Law, Ethics and Philosophy* 1: 143–160. <https://raco.cat/index.php/LEAP/article/view/294786>

Donaldson, Sue and Will Kymlicka. 2016a. "Locating Animals in Political Philosophy." *Philosophy Compass* 11 (11): 692–701. <https://doi.org/10.1111/pch3.12365>

Donaldson, Sue and Will Kymlicka. 2016b. "Between Wildness and Domestication: Rethinking Categories and Boundaries in Response to Animal Agency." *Animal Ethics in the Age of Humans: Blurring Boundaries of Human-animal Relationships.* Bernice Bovenkerk and Jozef Keulartz, eds. Springer. 225–239.

Duncanson, L., et al. 2023. "The Effectiveness of Global Protected Areas for Climate Change Mitigation." *Nature Communications* 14: 2908. <https://doi.org/10.1038/s41467-023-38073-9>

Eckersley, Robyn. 2004. *The Green State: Rethinking Democracy and Sovereignty.* Cambridge, MA: MIT P.

Ecology Trust, The. 2023. <https://www.ecologytrust.org/>

Einhorn, Catrin. 2024. "The Planet Needs Solar Power. Can We Build It Without Harming Nature?" *The New York Times*, 11 February. <https://www.nytimes.com/interactive/2024/02/11/climate/climate-change-wildlife-solar.html>

Eiseley, Loren. 1961. *Darwin's Century: Evolution and the Men Who Discovered It.* NY: Anchor Books.

Eisen, Michael B. and Patrick O. Brown. 2022. "Rapid Global Phaseout of Animal Agriculture Has the Potential to Stabilize Greenhouse Gas Levels for 30 Years and Offset 68 Percent of CO2 Emissions This Century." *Plos Climate* 1 (2): e0000010. <https://doi.org/10.1371/journal.pclm.0000010>

Elbert, Wolfgang, et al. 2012. "Contribution of Cryptogamic Covers to the Global Cycles of Carbon and Nitrogen." *Nature Geoscience*. <https://doi.org/10.1038/NGE01486>

Elliot, Robert. 1982. "Faking Nature." *Inquiry* 25 (1): 81–93.

Eshaghi Rad, Javad. 2018. "Effects of Anthropogenic Disturbance on Plant Composition, Plant Diversity and Soil Properties in Oak Forests, Iran." *Journal of Forest Science* 64 (8): 358–370. <https://doi.org/10.17221/13/2018-JFS>

Estrada-Villegas, Sergio, et al. 2022. "Animal Seed Dispersal Recovery During Passive Restoration in a Forested Landscape." *Philosophical Transactions of the Royal Society B* 378: 20210076. <https://doi.org/10.1098/rstb.2021.0076>

EU Parliament News. 2024. "Nature Restoration: Parliament adopts law to restore 20% of EU's land and sea." <https://www.europarl.europa.eu/news/en/press-room/20240223IPR18078/nature-restoration-parliament-adopts-law-to-restore-20-of-eu-s-land-and-sea>

European Court of Auditors. 2021. *EU Funding for Biodiversity and Climate Change in EU Forests.*

Evernden, Neil. 1993. *The Natural Alien: Humankind and Environment.* Second edition. Toronto: U Toronto P.

Fa, Julia E., et al. 2021. "Importance of Indigenous Peoples' Lands for the Conservation of Intact Forest Landscapes." *Frontiers in Ecology and the Environment* 18 (3): 135–140. <https://doi.org/10.1002/fee.2148>

FAO. 2020. *Global Forest Resources Assessment.* Rome. <https://doi.org/10.4060/ca8753en>

Farrelly, Colin. 2004. *An Introduction to Contemporary Political Philosophy.* London: Sage.

Feigin, Svetlana V., et al. 2023. "Proposed Solutions to Anthropogenic Climate Change: A Systematic Literature Review and a New Way Forward." *Heliyon* 9: e20544. <https://doi.org/10.1016/j.heliyon.2023.e20544>

Feuer, Lewis S., ed. *Karl Marx and Friedrich Engels: Basic Writings on Politics and Philosophy.* NY: Anchor Books, 1959.

Flyn, Cal. 2022. *Islands of Abandonment: Nature Rebounding in the Post Human Landscape.* NY: Penguin Books.

Foucault, Michel. 1991. *The Foucault Reader.* Paul Rabinow, ed. London: Penguin Books.

Fouts, Roger. 2001. "Darwinian Reflections on Our Fellow Apes." *Great Apes and Humans: The Ethics of Coexistence.* Benjamin B. Beck, et al., eds. Washington, DC: Smithsonian Institution Press. 191–211.

Fraser, Caroline. 2009. *Rewilding the World: Dispatches from the Conservation Evolution.* NY: Metropolitan Books.

Fraser, Erin, et al. 2006. "Carbohydrate Transfer Through Root Grafts to Support Shaded Trees." *Tree Physiology* 26: 1019–1023. <https://doi.org/10.1093/treephys/26.8.1019>

Friedman, Milton. 1970. "The Social Responsibility of Business is to Increase Its Profits." *The New York Times*, 13 September.

Gable, Thomas D., et al. 2023. "Wolves Alter the Trajectory of Forests by Shaping the Central Place Foraging Behaviour of an Ecosystem Engineer." *Proceedings of the Royal Society B* 290: 20231377. <https://doi.org/10.1098/rspb.2023.1377>

Garavito, Cesar Rodriguez. 2011. "Ethnicity.gov: Global Governance, Indigenous Peoples, and the Right to Prior Consultation in Social Minefields." *Indiana Journal of Global Legal Studies* 18 (1): 263–305.

Garnett, Stephen T., et al. 2018. "A Spatial Overview of the Global Importance of Indigenous Lands for Conservation." *Nature Sustainability* 1: 369–374. <https://doi.org/10.1038/s41893-018-0100-6>

Garthwaite, Josie. 2021. "Soils or Plants Will Absorb More CO2 as Carbon Levels Rise—But Not Both, Standford Study Finds." *Stanford News* 24 March. <https://news.stanford.edu/stories/2021/03/one-earths-biggest-carbon-sinks-overestimated>

Gatti, Luciana V., et al. 2021. "Amazonia as a Carbon Source Linked to Deforestation and Climate Change." *Nature* 595 (7867): 388–393. <https://doi.org/10.1038/s41586-021-03629-6>

Gauci, Vincent, et al. 2024. "Global Atmospheric Methane Uptake by Upland Tree Woody Surfaces." *Nature* 631: 796–800. <https://doi.org/10.1038/s41586-024-07592-w>

Geist, Helmut J. and Eric F. Lambin. 2002. "Proximate Causes and Underlying Driving Forces of Tropical Deforestation." *BioScience* 52 (2): 143–150.

Geremia, Chris, et al. 2019. "Migratory Bison Engineer the Green Wave." *PNAS* 116 (51): 25707–25713. <https://doi.org/10.1073/pnas.1913783116>

Gigliotti, Carol. 2022. *The Creative Lives of Animals*. NY: NYU Press.

Gilbert, Neil A., et al. 2022. "Human Disturbance Compresses the Spatiotemporal Niche." *PNAS* 119 (52): e2206339119. <https://doi.org/10.1073/pnas.2206339119>

Gilhen-Baker, et al. 2022. "Old Growth Forests and Large Old Trees as Critical Organisms Connecting Ecosystems and Human Health." *Environmental Chemistry Letters* 20: 1529–1538. <https://doi.org/10.1007/s10311-021-01372-y>

Global Forest Watch. <https://www.globalforestwatch.org/>

Gonzalez, Patrick. 2023a. "Natural Carbon Solutions Contribute to Halting Climate Change." *Parks Stewardship Forum* 39 (2). <https://doi.org/10.5070/P539260953>

Gonzalez, Patrick. 2023b. "Wildfire, Climate Change, Forest Resilience, and Carbon Solutions." *Parks Stewardship Forum* 39 (3). <https://doi.org/10.5070/P539362026>

Goodall, Jane. 2001. "Foreword." *Great Apes and Humans: The Ethics of Coexistence*. Benjamin B. Beck, et al., eds. Washington, DC: Smithsonian Institution Press. xiii–xxiv.

Grimsley, Ronald. 1967. "Jean-Jacques Rousseau." *Encyclopedia of Philosophy*. Paul Edwards, ed. NY: Macmillan. Volume 7: 218–225.

Guardian, The. 2022. "China's 26-storey Pig Skyscraper Ready to Slaughter 1 Million Pigs a Year." 25 November. <https://www.theguardian.com/environment/2022/nov/25/chinas-26-storey-pig-skyscraper-ready-to-produce-1-million-pigs-a-year>

Gudorf, Christine E. and James E. Huchingson. 2003. *Boundaries: A Casebook in Environmental Ethics*. Washington, DC: Georgetown UP.

Hadas, Moses. 1950. *A History of Greek Literature*. NY: Columbia UP.

Hadley, John. 2005. "Nonhuman Animal Property: Reconciling Environmentalism and Animal Rights." *Journal of Social Philosophy* 36 (3): 305–315. <http://dx.doi.org/10.1111/j.1467-9833.2005.00277.x>

Hajime, Nakamura. 1992. "The Idea of Nature in the East in Comparison with the West." *GeoJournal* 26:113–128. <https://doi.org/10.1007/BF00241205>

Hall, Donald E. 2001. *Literary and Cultural Theory*. Boston: Houghton Mifflin.

Hall, Shannon. 2015. "Exxon Knew About Climate Change Almost 40 Years Ago." *Scientific American* 26 (10).

Hannon, Elizabeth and Tim Lewens. 2014. "Cultural Evolution." *The Princeton Guide to Evolution*. Jonathn B. Losos, ed. in chief. Princeton: PUP. 795–800.

Hansen, Matthew C., et al. 2010. "Quantification of Global Gross Forest Cover Loss." *PNAS* 107 (19): 8650–8655. <www.pnas.org/cgi/doi/10.1073/pnas.0912668107>

Hardin, Garrett. 1968. "The Tragedy of the Commons." *Science* New Series 162 (3859): 1243–1248.

Hari, Johann. 2020. "The Wrong Kind of Green." *The Nation* 290 (11): 11–19.

Hart, P. Sol and Erik C. Nisbet. 2011. "Boomerang Effects in Science Communication." *Communication Research* 20 (10): 1–23. <https://doi.org/10.1177/0093650211416646>

Hassan, Aumyo and Sarah J. Barber. 2021. "The Effects of Repetition Frequency on the Illusory Truth Effect." *Cognitive Research* 6: 38. <https://doi.org/10.1186/s41235-021-00301-5>

Hawkins, Sally, et al. 2023. "Introduction: What is Rewilding?" *Routledge Handbook of Rewilding*. Sally Hawkins, et al., eds. London: Routledge. 3–10.

Hecht, Susanna B., et al. 2016. "Trees Have Already Been Invented: Carbon in Woodlands." *Collabra* 2 (1): 1–34. <http://dx.doi.org/10.1525/collabra.69>

Heilbroner, Robert L. 1961. *The Worldly Philosophers*. Revised ed. NY: Simon and Schuster.

Henrich, Joseph. 2016. *The Secret of Our Success*. Princeton: PUP.

Heinrich, Martin and Thom Tillis. 2023. "Recovering America's Wildlife Act." U.S. Senate, proposed bill, 118th Congress, 1st session.

Hibbett, David S. 2014. "Major Events in the Evolution of Fungi." *The Princeton Guide to Evolution*. Jonathan B. Losos, ed. in chief. Princeton: PUP. 152–158.

Hickman, Caroline, et al. 2021. "Climate Anxiety in Children and Young People and Beliefs About Government Responses to Climate Change: A Global Survey." *Lancet Planet Health* 5: e863–73. <https://doi.org/10.1016/S2542-5196(21)00278-3>

Hobbes, Thomas. 1952. *Leviathan, or Matter, Form, and Power of a Commonwealth Ecclesiastical and Civil.* Robert Hutchins, ed. Chicago: Britannica. (Original work published 1651)

Hobbs, N. Thompson, et al. 2024. "Does Restoring Apex Predators to Food Webs Restore Ecosystems? Large Carnivores in Yellowstone as a Model System." *Ecological Monographs* e1598. <https://doi.org/10.1002/ecm.1598>

Hoffman, Andrew J. 2015. *How Culture Shapes the Climate Change Debate.* Stanford: Stanford UP.

Holt, Flora Lu. 2005. "The Catch-22 of Conservation: Indigenous Peoples, Biologists, and Cultural Change." *Human Ecology* 33 (2): 199–215. <https://doi.org/10.1007/s10745-005-2432-X>

Holt, Robert D. 2014. "Evolution of the Ecological Niche." *The Princeton Guide to Evolution.* Jonathan B. Losos, ed. in chief. Princeton: PUP. 288–297.

Horkheimer, Max and Theodor W. Adorno. 2002. *Dialectic of Enlightenment: Philosophical Fragments.* Gunzelim Schmid Noerr, ed. Edmund Jephcott, trans. Stanford: Stanford UP. (Original work published 1944)

Houghton, Walter E. 1957. *The Victorian Frame of Mind, 1830–1870.* New Haven: Yale UP.

Hutchinson, G. E. 1959. "Homage to Santa Rosalia or Why Are There So Many Kinds of Animals?" *The American Naturalist* 93 (870): 145–159.

Ibrahim, Hindou Oumarou, et al. 2025. "Joint Letter by Chairpersons of UNPFII, Expert Mechanism and Special Rapporteur on the Rights of Indigenous Peoples." U.N. Dept. of Economic and Social Affairs. Final JOINT LETTER 26 Feb (003).pdf

James, William. 1961. *The Varieties of Religious Experience.* NY: Collier Books.

Jamieson, Dale. 2018. "Animal Agency." *The Harvard Review of Philosophy* 25: 111–126. oi: 10.5840/harvardreview201892518

Jay, A. K., et al. 2023. "Overview: Understanding Risks, Impacts, and Responses." *Fifth National Climate Assessment.* A. R. Crimmins, et al., eds. Washington, DC: U.S. Global Change Research Program. <https://doi.org/10.7930/NCA5.2023.CH1>

Jones, Clive G., et al. 1994. "Organisms as Ecosystem Engineers." *Oikos* 69 (3): 373–386.

Jones, Clive G., et al. 1997. "Positive and Negative Effects of Organisms as Physical Ecosystem Engineers." *Ecology* 78. 1946–1957.

Joy, Yannick and Andreas De Block. 2025. "Our Not-so-natural Connection to Nature." *Trends in Ecology and Evolution.* <https://doi.org/10.1016/j.tree.2025.01.003>

Kallis, Giorgos and Hug March. 2015. "Imaginaries of Hope: The Utopianism of Degrowth." *Annals of the Association of American Geographers* 105 (2): 360–368. <https://doi.org/10.1080/00045608.2014.973803>

Kant, Immanuel. 1996. *The Metaphysics of Morals*. Mary Gregor, trans. Cambridge: CUP. (Original work published 1797)

Kaufmann, Walter. 1974. *Nietzsche: Philosopher, Psychologist, Antichrist*. Princeton: PUP.

Keuepper, Barbara, et al. 2019. "Feed and Livestock in Brazil, China, EU Consume Most Cerrado Soy." *Chain Reaction Research*. <https://chainreactionresearch.com/wp-content/uploads/2019/12/Feed-and-Livestock-in-Brazil-China-EU-Consume-Most-Cerrado-Soy-.pdf>

Kim, Rakhyun E. 2021. "Taming Gaia 2.0: Earth System Law in the Ruptured Anthropocene." *The Anthropocene Review* 9 (3): 411–424. <https://doi.org/10.1177/20530196211026721>

Klein, Naomi. 2016. "Let Them Drown: The Violence of Othering in a Warming World." *London Review of Books* 38 (11). <https://www.lrb.co.uk/the-paper/v38/n11/naomi-klein/let-them-drown>

Kohda, M., et al. 2019. "If a Fish Can Pass the Mark Test, What Are the Implications for Consciousness and Self-awareness Testing in Animals?" *PLOS Biology* 17 (2): e3000021. <https://doi.org/10.1371/journal.pbio.3000021>

Kohn, Eduardo. 2013. *How Forests Think: Toward an Anthropology Beyond the Human*. Berkeley: U California P.

Kopenawa, Davi and Bruce Albert. 2013. *The Falling Sky: Words of a Yanomami Shaman*. Nicholas Elliott and Alison Dundy, trans. Cambridge, MA: Belknap P.

Korsgaard, Christine M. 2018. *Fellow Creatures: Our Obligations to Other Animals*. Oxford: OUP.

Kret, Mariska E., et al. 2022. "My Fear is Not, and Never Will Be, Your Fear: On Emotions and Feelings in Animals." *Affective Science*. <https://doi.org/10.1007/s42761-021-00099-x>

Kropotkin, Peter. 1902. *Mutual Aid: A Factor of Evolution*. NY: McClure Phillips.

Kuehn, Manfred. 2001. *Kant: A Biography*. Cambridge: CUP.

Kuhn, Thomas S. 2012. *The Structure of Scientific Revolutions*. Fourth edition. Chicago: U Chicago P.

Kumar, Victor and Richmond Campbell. 2022. *A Better Ape: The Evolution of the Moral Mind and How it Made Us Human*. NY: Oxford UP.

Kymlicka, Will. 2002. *Contemporary Political Philosophy*. Second edition. Oxford: OUP.

Kymlicka, Will and Sue Donaldson. 2014. "Animals and the Frontiers of Citizenship." *Oxford Journal of Legal Studies*. 34 (2): 201–219. <https://doi.org/10.1093/ojls/gqu001>

Kymlicka, Will and Sue Donaldson. 2015. "Animal Rights and Aboriginal Rights." *Canadian Perspectives on Animals and the Law*. Toronto: Irwin Law. 159–186.

Ladwig, Bernd. 2015. "Against Wild Animal Sovereignty: An Interest-based Critique of Zoopolis." *The Journal of Political Philosophy* 23 (3): 282–301.

Laigle, Idaline, et al. 2021. "Direct and Indirect Effects of Forest Anthropogenic Disturbance on Above and Below Ground Communities and Litter Decomposition." *Ecosystems* 24: 1716–1737. <https://doi.org/10.1007/s10021-021-00613-z>

Leach, Stefan, et al. 2022. "Changing Minds About Minds: Evidence that People are Too Sceptical About Animal Sentience." *Cognition* 230: 105263. <https://doi.org/10.1016/j.cognition.2022.105263>

Ledger, Sophie E. H., et al. 2022. *Wildlife Comeback in Europe: Opportunities and Challenges for Species Recovery*. London: Zoological Society of London.

Lee, Hoesung, et al. 2023. *IPCC Climate Change 2023, Synthesis Report*. <https://doi.org/10.59327/IPCC/AR6-9789291691647.001>

Lee, Yumi and Yeon-A Son. 2016. "Analysis and Comparison of Views of Nature Between East Asia and the Western World and Its Meaning." *Journal of the Korean Association for Science Education* 36.3: 485–498. <https://doi.org/10.14697/jkase.2016.36.3.0485>

Leiserowitz, Anthony, et al. 2023. *International Public Opinion on Climate Change*. Yale Program on Climate Change Communication. <https://climatecommunication.yale.edu/publications/international-public-opinion-on-climate-change-2023/>

Leopold, Aldo. 1949. "The Land Ethic." *A Sand County Almanac*. London: Oxford UP. 201–226.

Lewandowsky, Stephen, et al. 2012. "Misinformation and Its Correction: Continued Influence and Successful Debiasing." *Psychological Science in the Public Interest* 13 (3): 106–131. <https://doi.org/10.1177/1529100612451018>

Lindo, Zoë and Jonathan A. Whiteley. 2011. "Old Trees Contribute Bio-available Nitrogen Through Canopy Bryophytes." *Plant Soil* 342: 141–148. <https://doi.org/10.1007/s11104-010-0678-6>

Locke, John. 1952. *A Letter Concerning Toleration* and *The Second Essay Concerning the True Original Extent and End of Civil Government*. Robert Hutchins, ed. Chicago: Britannica. (Original works published 1689 and 1690)

Lovejoy, Thomas E. and Carlos Nobre. 2018. "Amazon Tipping Point." *Science Advances* 4 (2): eaat2340. <https://doi.org/10.1126/sciadv.aat2340>

Lovelock, James. 1979. *Gaia: A New Look at Life on Earth*. Oxford: OUP.

Lovelock, James. 2000. Preface. *Gaia: A New Look at Life on Earth*. Oxford: OUP.

Lovelock, James. 2005. Foreword. *The Essential Mary Midgley*. David Midgley, ed. London: Routledge. vii–ix.

Luke, Timothy W. 1999. *Capitalism, Democracy, and Ecology: Departing From Marx*. Urbana, IL: U Illinois P.

Macedo, Marcia N., et al. 2012. "Decoupling of Deforestation and Soy Production in the Southern Amazon During the Late 2000s." *PNAS* 109 (4): 1341–1346. <www.pnas.org/cgi/doi/10.1073/pnas.1111374109>

Machovina, Brian, et al. 2015. "Biodiversity Conservation: The Key is Reducing Meat Consumption." *Science of the Total Environment* 536: 419–431. <https://doi.org/10.1016/j.scitotenv.2015.07.022>

Mack, Eric. 2022. "Robert Nozick's Political Philosophy." *Stanford Encyclopedia of Philosophy*. <https://plato.stanford.edu/entries/nozick-political/>

Mansbridge, Jane. 2013. "The Role of the State in Governing the Commons." *Environmental Science and Policy* 36: 8–10. <http://dx.doi.org/10.1016/j.envsci.2013.07.006>

Marcus, Gary. 2022. "AI Platforms Like ChatGPT Are Easy to Use But Also Potentially Dangerous." *Scientific American* 19 December. <https://www.scientificamerican.com/article/ai-platforms-like-chatgpt-are-easy-to-use-but-also-potentially-dangerous/>

Margulis, Lynn. 1998. *Symbiotic Planet: A New Look at Evolution*. NY: Basic Books.

Marx, Karl. 1952. *Capital*. Friedrich Engels, ed. Chicago: Britannica. (Original work published 1867)

Mayer, Claude-Hélène. 2021. "Understanding Wildlife Crime from Eco-Existential and African Perspectives." *International Journal of Environmental Research and Public Health* 18 (11675). <https://doi.org/10.3390/ijerph182111675>

Mayr, Ernst. 1991. *One Long Argument: Charles Darwin and the Genesis of Modern Evolutionary Thought*. Cambridge, MA: Harvard UP.

Mayr, Ernst. 1997. *This is Biology: The Science of the Living World*. Cambridge, MA: Belknap P.

McCabe, J. Terrence. 1990. "Turkana Pastoralism: A Case Against the Tragedy of the Commons." *Human Ecology* 18: 81–103. <https://doi.org.10.1007/BF00889073>

McArthur, Robert H. and E. O. Wilson. 1967. *The Theory of Island Biogeography*. Princeton: PUP.

McGinley, Kathleen, et al. 2012. "Regulating the Sustainability of Forest Management in the Americas: Cross-country Comparisons of Forest Legislation." *Forests* 3: 467–505. <https://doi.org/10.3390/f3030467>

McInnes, Neil. 1967. "Karl Marx." *Encyclopedia of Philosophy*. Paul Edwards, ed. NY: Macmillan. Volume 5: 171–173.

McNeill, William H. 1976. *Plagues and Peoples*. Garden City, NY: Anchor/ Doubleday.

Meijer, Eva. 2019. *When Animals Speak: Toward an Interspecies Democracy*. NY: NYU Press.

Midgley, Mary. 2005. *The Essential Mary Midgley*. David Midgley, ed. London: Routledge.

Milburn, Josh. 2017. "Robert Nozick on Nonhuman Animals: Rights, Value and the Meaning of Life." *Ethical and Political Approaches to Nonhuman Animal Issues*. A. Woodhall and G. Garmendiada Trindade, eds. Cham: Palgrave Macmillan. 97–120.

Milburn, Josh. 2020. "Animal Sovereignty Theory." *Global Encyclopedia of Territorial Rights*. M. Kocsis, ed. Cham: Springer. 1–4.

Mill, John Stuart. *On Liberty*. Stefan Collini, ed. Cambridge: CUP, 1989. (Original work published 1859)

Miller, David. 2003. *Political Philosophy*. Oxford: OUP.

Mo, Lidong, et al. 2023. "Integrated Global Assessment of the Natural Forest Carbon Potential." *Nature*. <https://doi.org/10.1038/s41586-023-06723-z>

Monsó, Susana, et al. 2018. "Animal Morality: What it Means and Why it Matters." *Journal of Ethics* 22 (3): 283–310. <https://doi.org/10.1007/s10892-018-9275-3>

Moore, Thomas Gale. 1997. "Happiness is a Warm Planet." *The Wall Street Journal*, 7 October.

Morris, Viveca and Jennifer Jacquet. 2024. "The Animal Agriculture Industry, US Universities, and the Obstruction of Climate Understanding and Policy." *Climate Change* 177: 41. <https://doi.org/10.1007/s10584-024-03690-w>

Mortelliti, Alessio and Allison M. Brehm. 2020. "Environmental Heterogeneity and Population Density Affect the Functional Diversity of Personality Traits in Small Mammal Populations." *Proceedings of the Royal Society B* 287: 20201713. <https://doi.org/10.1098/rspb.2020.1713>

Moynihan, Daniel Patrick. 1969. Memorandum from White House to John Ehrlichman Re Fossil Fuels and CO2 Increase.

Nadasy, Paul. 2016. "First Nations, Citizenship and Animals, or Why Northern Indigenous People Might Not Want to Live in a Zoopolis." *Canadian Journal of Political Science* 49 (1): 1–20<. https://doi.org/10.1017/S0008423915001079>

Naess, Arne. 1973. "The Shallow and the Deep, Long-Range Ecology Movement." *Inquiry* 16 (1): 95–100.

Nagel, Thomas. 2013. Foreword. *Anarchy, State, and Utopia*. Robert Nozick. NY: Basic Books.

Nepstad, Daniel C., et al. 2008. "Interactions Among Amazon Land Use, Forests, and Climate: Prospects for a Near-term Forest Tipping Point." *Philosophical*

Transactions of the Royal Society B 363: 1737–1746<. https://doi.org/10.1098/rstb.2007.0036>

Niklas, Karl J. 2016. *Plant Evolution: An Introduction to the History of Life*. Chicago: U Chicago P.

NOAA. 2023. "How Much Oxygen Comes From the Ocean?" National Oceanic and Atmospheric Administration. <https://oceanservice.noaa.gov/facts/ocean-oxygen.html#:~:text=At>

Norton Marcy. 2024. *The Tame and the Wild: People and Animals After 1492*. Harvard UP.

Noss, R. F., et al. 2021. *Science Needs of Southeastern Grassland Species of Conservation Concern—A Framework for Species Status Assessments: U.S. Geological Survey Open-File Report 2021–1047.* <https://doi.org/10.3133/ofr20211047>

Nozick, Robert. 1974. *Anarchy, State, and Utopia*. NY: Basic Books.

Nussbaum, Martha C. 2022. *Justice for Animals: Our Collective Responsibility*. NY: Simon and Schuster.

NYU. 2024. The New York Declaration on Animal Consciousness. <https://sites.google.com/nyu.edu/nydeclaration/declaration>

Ogden, Laura A., et al. 2013. "Animals, Plants, People, and Things." *Environment and Science: Advances in Research* 4: 5–24<. https://doi.org/10.3167/ares.2013.040102>

Oreskes, Naomi. 2018. "The Scientific Consensus on Climate Change: How Do We Know We're Not Wrong?" *Climate Modelling*. Elisabeth A. Lloyd and Eric Winsberg, eds. Cham Switzerland: Palgrave. 31–64.

Osuri, Anand M., et al. 2020. "Greater Stability of Carbon Capture in Species-rich Natural Forests Compared to Species-poor Plantations." *Environmental Research Letters* 15: 034011. <https://doi.org/10.1088/1748-9326/ab5f75>

Packe, Michael St. John. *The Life of John Stuart Mill*. NY: Capricorn Books, 1970.

Peters, R. S. 1967. "Thomas Hobbes." *Encyclopedia of Philosophy*. Paul Edwards, ed. NY: Macmillan. Volume 4: 30–46.

Piccolo, John J., et al. 2022. "'Nature's Contributions to People' and People's Moral Obligations to Nature." *Biological Conservation* 270: 109572. <https://doi.org/10.1016/j.biocon.2022.109572>

Plumwood, Val. 2001. *Environmental Culture: The Ecological Crisis of Reason*. London: Routledge.

Pörtner, Hans-Otto, et al., eds. 2022. *Climate Change 2022: Sixth Assessment Report: Impacts, Adaptation and Vulnerability*. Cambridge: CUP. <https://doi.org/10.1017/9781009325844>

Preston, Diana. 2022. *The Evolution of Charles Darwin*. NY: Grove Press.

Przybylska, Maria Stefania, et al. 2024. "Can Plants Build Their Niche Through Modulation of Soil Microbial Activities Linked With Nitrogen Cycling? A

Test With *Arabidopsis thaliana*." *New Phytologist*. <https://doi.org/10.1111/nph.19870>

Pulido, Laura. 2017. "Geographies of Race and Ethnicity II: Environmental Racism, Racial Capitalism and State-sanctioned Violence." *Progress in Human Geography* 41 (4): 524–533. <https://doi.org/10.1177/0309132516646495>

Raskin, Paul, et al. 2002. *Great Transition: The Promise and Lure of the Times Ahead*. Boston: Stockholm Environment Institute.

Raskin, Paul. 2021. *Journey to Earthland: The Great Transition to Planetary Civilization*. Cambridge, MA: Tellus Institute.

Rawlence, Ben. 2022. *The Treeline: The Last Forest and the Future of Life on Earth*. NY: St. Martin's.

Rawls, John. 1999. *A Theory of Justice*. Revised edition. Cambridge, MA: Belknap/Harvard UP.

Reed, Stanley. 2024. "Cement is a Big Polluter. A Plant in Norway Hopes to Clean It Up." *The New York Times*, 6 October. <https://www.nytimes.com/2024/10/06/business/cement-pollution-carbon-capture-norway.html>

Rees, William E. 2008. "Confounding Integrity: Humanity as Dissipative Structure." *Reconciling Human Existence with Ecological Integrity*. Laura Westra, et al., eds. London: Earthscan. 293–302.

Reid, John W. and Thomas E. Lovejoy. 2022. *Ever Green: Saving Big Forests to Save the Planet*. NY: W. W. Norton.

Reid Joshua. 2021. "Replacing Rights with Indigenous Relationality to Reclaim Homelands." *Bridging Cultural Concepts of Nature: Indigenous People and Protected Spaces of Nature*. Boyd Cothran and Saara Kekki, eds. Helsinki: Helsinki UP. 261–306.

Regan, Tom. 2004. *The Case for Animal Rights*. Updated edition. Berkeley, CA: U California P.

Rénique, Gerardo. 2009. "Law of the Jungle in Peru: Indigenous Amazonian Uprising Against Neoliberalism." *Socialism and Democracy* 23 (3): 117–135. <https://doi.org/10.1080/08854300903290835>

Report of The Environmental Pollution Panel, President's Science Advisory Committee, The White House. 1965.

Rijksen, Herman D. 2001. "The Orangutan and the Conservation Battle in Indonesia." *Great Apes and Humans: The Ethics of Coexistence*. Benjamin B. Beck, et al., eds. Washington, DC: Smithsonian Institution Press. 57–70.

Ritchie, Hannah. 2017, revised 2019. "Meat and Dairy Production." *Our World in Data*. <https://ourworldindata.org/meat-production>

Ritchie, Hannah. 2022. "Wild Mammals Make Up Only a Few Percent of The World's Mammals." *Our World in Data*. <https://ourworldindata.org/wild-mammals-birds-biomass>

Ritchie, Hannah and Max Roser. 2021. "Forests and Deforestation." *Our World in Data*. <https://ourworldindata.org/forests-and-deforestation>

Robinson, Guy M. 2024. "Global Sustainable Agriculture and Land Management Systems." *Geography and Sustainability* 5: 637–646. <https://doi.org/10.1016/j.geosus.2024.09.001>

Rockström, John and Mattias Klum. 2015. *Big World Small Planet: Abundance Within Planetary Boundaries*. New Haven, CT: Yale UP.

Ross, W. D. 1998. *Ethics: The Classic Readings*. David E. Cooper, ed. Oxford: Blackwell. 246–261.

Rousseau, Jean-Jacques. 1952. *The Social Contract*. G. D. H. Cole, trans. Robert Hutchins, ed. Chicago: Britannica. (Original work published 1762)

Ruddiman, Willaim F. 2003. "The Anthropogenic Greenhouse Era Began Thousands of Years Ago." *Climate Change* 61: 261–293.

Russell, Bertrand. 1952. *The Problems of Philosophy*. Mortimer J. Adler, ed. Chicago: Britannica. (Original work published 1912)

Sagan, Dorion and Lynn Margulis. 2023. *Gaia and Philosophy*. Great Britain: Ignota Books.

Salati, Eneas, et al. 1979. "Recycling of Water in the Amazon Basin: An Isotopic Study." *Water Resources Research* 15 (5): 1250–1258. <https://doi.org/10.1029/WR015i005p01250>

Scanlon, T. M. 1982. "Contractualism and Utilitarianism." *Utilitarianism and Beyond*. Amartya Sen and Bernard Williams, eds. Cambridge: CUP. 102–128.

Schewenius, Maria. 2014. "Cities and Biodiversity Outlook." Stockholm Resilience Centre. <https://www.stockholmresilience.org/download/18.10119fc11455d3c557d3dc38/1459560235045/SRC_brief_1_CBO.pdf>

Schlossberg, David. 2007. *Defining Environmental Justice*. Oxford: OUP.

Schmitt, Richard. 2009. *An Introduction to Social and Political Philosophy*. Lanham, MD: Rowman and Littlefield.

Schneewind, J. B. 1967. "John Stuart Mill." *Encyclopedia of Philosophy*. Paul Edwards, ed. NY: Macmillan. Volume 5: 314–323.

Schweiger, Stefan and Ulrike Cress. 2019. "Attitude Confidence and Source Credibility in Information Foraging with Social Tags." *Plos One* 14 (1): e0210423. <https://doi.org/10.1371/journal.pone.0210423>

Searchinger, Timothy, et al. 2008. "Use of U.S. Croplands for Biofuels Increases Greenhouse Gases Through Emissions From Land-use Change." *Science* 319 (5867): 1238–1240. <https://doi.org/10.1126/science.1151861>

Sebo, Jeff. 2023. "Kantianism for Humans, Utilitarianism for Nonhumans? Yes and No." *Philosophical Studies* 180: 1211–1230. <https://doi.org/10.1007/s11098-022-01835-0>

SEI Climate Analytics, E3G, IISD, and UNEP. 2023. *The Production Gap: Phasing Down or Phasing Up? Top Fossil Fuel Producers Plan Even More Extraction Despite Climate Promises*. Stockholm Environment Institute, Climate Analytics, E3G, International Institute for Sustainable Development and United Nations Environment Programme. <https://doi.org/10.51414/sei2023.050>

Serres, Michel. 1995. *The Natural Contract*. Elizabeth MacArthur and William Paulson, trans. Ann Arbor, MI: U Michigan P. (Original work published 1990)

Serres, Michel. 2011. Malfeasance: Appropriation Through Pollution? Anne-Marie Feenberg-Dibon, trans. Stanford: Stanford UP. (Original work published 2008)

Silveira, Maria, et al. 2012. "Carbon Sequestration in Grazing Land Ecosystems." Document SL373, Department of Soil and Water Sciences, Univ. of Florida/ IFAS Extension.

Simard, Suzanne, et al. 1997. "Net Transfer of Carbon Between Ectomycorrhizal Tree Species in the Field." *Nature 388*: 579–582. <https://doi.org/10.1038/41557>

Simard, Suzanne. 2021. *Finding the Mother Tree: Discovering the Wisdom of the Forest*. NY: Vintage Books.

Simmons, A. John. 2008. *Political Philosophy*. NY: Oxford.

Singer, Peter. 2009. *Animal Liberation*. Updated edition. NY: Ecco. (Original work published 1975)

Smith, Adam. 1952. *An Inquiry into the Nature and Causes of the Wealth of Nations*. Robert Hutchins, ed. Chicago: Britannica. (Original work published 1776)

Smith, Mick. 2011. *Against Ecological Sovereignty: Ethics, Biopolitics, and Saving the Natural World*. Minneapolis: U of Minnesota P.

Smith, Steven B. 2012. *Political Philosophy*. New Haven: Yale UP.

Song, Yuan Yuan, et al. 2015. "Defoliation of Interior Douglas-fir Elicits Carbon Transfer and Stress Signalling to Ponderosa Pine Neighbors Through Ectomycorrhizal Networks." *Scientific Reports* 5: 8495. <https://doi.org/10.1038/srep08495>

Song, Xiao-Peng, et al. 2018. "Global Land Change 1982–2016." *Nature* 560 (7720): 639–643. <https://doi.org/10.1038/s41586-018-0411-9>

Song, Xiao-Peng, et al. 2021. "Massive Soybean Expansion in South America Since 2000 and Implications for Conservation." *Nature Sustainability*. <https://doi.org/10.1038/s41893-021-00729-z>

Soulé, Michael and Reed Noss. 1998. "Rewilding and Biodiversity: Complementary Goals for Continental Conservation." *Wild Earth*. 18–28.

Sprague, Elmer. 1967. "Adam Smith." *Encyclopedia of Philosophy*. Paul Edwards, ed. NY: Macmillan. Volume 7: 461–463.

Springmann, Marco, et al. 2018. "Options for Keeping the Food System Within Environmental Limits." *Nature 562*: 519–525. <https://doi.org/10.1038/s41586-018-0594-0>

Stäger, Felix, et al. 2023. “Disproportionately High Contributions of 60 Year Old Weapons-137Cs Explains the Persistence of Radioactive Contamination in Bavarian Wild Boars.” *Environmental Science and Technology*. <https://doi.org/10.1021/acs.est.3c03565>

Stoinski, Tara S., et al. 2001. “Captive Apes and Zoo Education.” *Great Apes and Humans: The Ethics of Coexistence*. Benjamin B. Beck, et al., eds. Washington, DC: Smithsonian Institution Press. 113–132.

Stone, Christopher D. 2010. *Should Trees Have Standing? Law, Morality, and the Environment*. Third ed. Oxford: OUP.

Strassmann, Joan E. and David C. Queller. 2014. “Cooperation and Conflict: Microbes to Humans.” *The Princeton Guide to Evolution*. Jonathan B. Losos, ed. in chief. Princeton: PUP. 671–676.

Strauss, Sharon Y. 2014. “Adaptation to the Biotic Environment.” *The Princeton Guide to Evolution*. Jonathan B. Losos, ed. in chief. Princeton: PUP. 298–303.

Stephenson, N. L., et al. 2014. “Rate of Tree Carbon Accumulation Increases Continuously with Tree Size.” *Nature* 507: 90–93. <https://doi.org/10.1038/nature12914>

Supran, G., et al. 2023. “Assessing ExxonMobil’s Global Warming Projections.” *Science* 379 (153): eabkoo63. <https://doi.org/10.1126/science.abk0063>

Tague, Gregory F. 2016. *Evolution and Human Culture*. Leiden: Brill/Rodopi.

Tague, Gregory F. 2020. *An Ape Ethic and the Question of Personhood*. Lanham, MD: Lexington Books.

Tague, Gregory F. 2022. *The Vegan Evolution*. London: Routledge.

Tankersley, Jim, et al. 2023. “The Clean Energy Future is Roiling Both Friends and Foes.” *The New York Times*, 12 August. <https://www.nytimes.com/interactive/2023/08/12/climate/wind-solar-clean-energy.html>

Tessler, Michael, et al. 2023. “Rewilding in Miniature: Suburban Meadows Can Improve Soil Microbial Biodiversity and Soil Health.” *Microbial Ecology* 85: 1077–1086. <https://doi.org/10.1007/s00248-023-02171-4>

Theurl, Michaela C., et al. 2020. “Food Systems in a Zero-deforestation World: Dietary Change is More Important Than Intensification for Climate Targets in 2050.” *Science of the Total Environment* 735: 139353. <https://doi.org/10.1016/j.scitotenv.2020.139353>

TOP. 2024. The Orangutan Project. “Secure, Restore and Patrol Orangutan Habitat.” Accessed 3 February. <https://www.theorangutanproject.org/projects-overview/secure-restore-and-patrol-orangutan-habitat/>

Trawick, Paul. 2003. “Against the Privatization of Water: An Indigenous Model for Improving Existing Laws and Successfully Governing the Commons.” *World Development* 31 (6): 977–996. <https://doi.org/10.1016/S0305-750X(03)00049-4>

Tyson, Alec and Brian Kennedy. 2023. "How Americans View Future Harms From Climate Change in Their Community and Around the U.S." Washington, D.C.: Pew Research Center.

Uexküll, Jakob von. 2010. *A Foray into the Worlds of Animals and Humans*. Joseph D. O'Neil, trans. Minneapolis: U Minnesota P.

Vahabava, Maryna, et al. 2022. *Circular Economy*. U.N. University for Peace. Rome, Italy.

Van Bael, Sunshine A., et al. 2003. "Birds Defend Trees from Herbivores in a Neotropical Forest Canopy." PNAS 100 (14): 8304–8307. <https://doi.org/10.1073/pnas.1431621100>

Wadiwel, Dinesh Joseph. 2015. *The War Against Animals*. Leiden: Brill.

Waggoner, Paul E. and Jesse H. Ausubel. 2001. "How Much Will Feeding More and Wealthier People Encroach on Forests?" *Population and Development Review* 27 (2): 239–257.

Wakabayashi, Daisuke and Claire Fu. 2023. "China's Bid to Improve Food Production? Giant Towers of Pigs." *The New York Times*, 25 November. <https://www.nytimes.com/2023/02/08/business/china-pork-farms.html>

Walker, Wayne S., et al. 2020. "The Role of Forest Conversion, Deregulation, and Disturbance in the Carbon Dynamics of Amazon Indigenous Territories and Protected Areas." *PNAS* 117 (6): 3015–3025. <www.pnas.org/cgi/doi/10.1073/pnas.1913321117>

Walker, Wayne S., et al. 2022. "The Global Potential for Increased Storage of Carbon on Land." *PNAS* 119 (23): e2111312119. <https://doi.org/10.1073/pnas.2111312119>

Walzer, Michael. 1983. *Spheres of Justice: A Defense of Pluralism and Equality*. NY: Basic Books.

Wang, Yue, et al. 2023. "Risk to Rely on Soil Carbon Sequestration to Offset Global Ruminant Emissions." *Nature Communications* 14: 7625. <https://doi.org/10.1038/s41467-023-43452-3>

Warren, Mary Anne. 2001. "The Moral Status of Great Apes." *Great Apes and Humans: The Ethics of Coexistence*. Benjamin B. Beck, et al., eds. Washington, DC: Smithsonian Institution Press. 313–328.

Watson, James E. M., et al. 2018. "Protect the Last of the Wild." *Nature* 563: 27–30. <https://doi.org/https://doi.org/10.1038/d41586-018-07183-6>

Wenar, Leif. 2021. "John Rawls." *Stanford Encyclopedia of Philosophy*. <https://plato.stanford.edu/entries/rawls/>

Westra, Evan, et al. 2024. "In Search of Animal Normativity: A Framework for Studying Social Norms in Non-human Animals." *Biological Reviews*. <https://doi.org/10.1111/brv.13056>

Whittington, Jesse, et al. 2022. "Towns and Trails Drive Carnivore Movement Behaviour, Resource Selection, and Connectivity." *Movement Ecology* 10: 17. <https://doi.org/10.1186/s40462-022-00318-5>

Wilkie, David S. 2001. "Bushmeat Trade in the Congo Basin." *Great Apes and Humans: The Ethics of Coexistence*. Benjamin B. Beck, et al., eds. Washington, DC: Smithsonian Institution Press. 86–109.

Williams, Brooke A., et al. 2024. "Global Potential for Natural Regeneration in Deforested Tropical Regions." *Nature*. <https://doi.org/10.1038/s41586-024-08106-4>

Wilson, E. O. 1984. *Biophilia: The Human Bond with Other Species*. Cambridge, MA: Harvard UP.

Wilson, E. O. 2016. *Half-Earth: Our Planet's Fight for Life*. NY: Liveright Publishing.

Wilson, E. O. 2002. Afterword. *Silent Spring*, Rachel Carson. Boston: Mariner Books. 357–363.

Winton, R. I. and Peter Garnsey. 1981. "Political Theory." *The Legacy of Greece*. M. I. Finley, ed. Oxford: Clarendon P. 37–64.

Wohlleben, Peter. 2016. *The Hidden Life of Trees: What They Feel, How They Communicate: Discoveries From a Secret World*. Jane Billinghurst, trans. London: William Collins.

Wohlleben, Peter. 2019. *The Secret Network of Nature*. Jane Billinghurst, trans. Vancouver: Greystone Books.

Wolff, Jonathan. 2006. *An Introduction to Political Philosophy*. Oxford: OUP.

Wood, Chelsea L., et al. 2023. "A Reconstruction of Parasite Burden Reveals One Century of Climate-associated Parasite Decline." *PNAS* 120 (3): e2211903120. <https://doi.org/10.1073/pnas.2211903120>

Wrangham, Richard W. 2001. "Moral Decisions About Wild Chimpanzees." *Great Apes and Humans: The Ethics of Coexistence*. Benjamin B. Beck, et al., eds. Washington, DC: Smithsonian Institution Press. 230–244.

Wuebbles, D. J., et al. 2017. "Executive Summary." *Climate Science Special Report: Fourth National Climate Assessment, Volume 1*. Washington: U.S. Global Change Research Program. <https://doi.org/10.7930/J0DJ5CTG>

Wulf, Andrea. 2015. *The Invention of Nature: Alexander Von Humboldt's New World*. NY: Alfred A. Knopf.

Xu, Xiaoming, et al. 2021. "Global Greenhouse Gas Emissions from Animal-based foods are Twice Those of Plant-based Foods." *Nature Food* 2: 724–732. <https://doi.org/10.1038/s43016-021-00358x>

Zhao, Wanying, et al. 2020. "Field Test and Economic Analysis of Energy-saving Renovation for an Old Nursery Pig Building in Beijing, China." *Applied Engineering in Agriculture* 36 (5): 619–628. <https://doi.org/10.1301/aea.13655>

Zonca, Vincent. 2023. *Lichens: Toward a Minimal Resistance*. Jody Gladding, trans. Cambridge: Polity Press.

Index

PLANTS AND ANIMALS
INTERDISCIPLINARY APPROACHES

Series Editors
Professor Susan McHugh, University of New England
Professor Patrícia Vieira, University of Coimbra

Plants and Animals: Interdisciplinary Approaches aims to publish scholarly work that addresses common challenges across the fields of plant and animal studies from interdisciplinary perspectives. The series welcomes monographs and edited collections that focus and reflect upon interactions of plants, animals, and humans in innovative ways. At a time of large-scale anthropogenic species extinction, there is a pressing need to promote scholarship that can help us envision more equitable and harmonious forms of coexistence on the planet. The series therefore encourages submissions explicitly geared to build bridges not only between plant and animal studies, but also leading-edge research on other forms of life or ways of being, including fungi, lichens, algae and other microorganisms, as well as scholarship on fantasy creatures, cryptids, semi-living beings, and even non-living forms of existence. The goal is to abolish an artificially compartmentalized view of the world in order to add to the ways of knowing that are beginning to grow through the interconnections between these related fields of study. Grounded in the humanities, *Plants and Animals* welcomes trans-disciplinary perspectives that engage with scholarship in the social sciences and in the natural sciences.

Vol. 1 Gregory F. Tague, Forest Sovereignty: Wildlife Sustainability and Ethics. 2025.

www.ingramcontent.com/pod-product-compliance
Ingram Content Group UK Ltd.
Pitfield, Milton Keynes, MK11 3LW, UK
UKHW020212290926
15170UKWH00001B/144

9 781803 749556